Electrical and Electronic Engineering Principles

A READY-REFERENCE GUIDE

to quantities, units, symbols, definitions, formulae and circuit diagram symbols

Compiled by
J.O. Bird B.Sc. (Hons), C.Eng. M.I.E.E., F.COLL.P.,
F.I.M.A., T.Eng., M.I.ELEC.I.E.

Longman
Scientific &
Technical

Longman Scientific & Technical,
Longman Group UK Limited,
Longman House, Burnt Mill, Harlow,
Essex CM20 2JE, England
and Associated Companies throughout the world.

AL (iss)

028854838

First published 1987

British Library Cataloguing in Publication Data

Bird, J.O.
  Electrical and electronic engineering principles:
  a ready reference guide to quantities, units,
  symbols, definitions, formulae and circuit diagram
  symbols.
  1. Electric engineering — Handbooks, manuals, etc.
  I. Title
  621.3′0212   TK151

ISBN 0-582-41399-0

Set in 10 pt Times by MHL Typesetting Ltd, Coventry

Produced by Longman Singapore Publishers (Pte) Ltd.
Printed in Singapore.

621.3

Contents

Contents

Preface

There are a large number of important formulae involved in the area of electrical and electronic engineering; also, the various quantity definitions, units and symbols are numerous.

The aim of this booklet is to present concisely a 'ready-reference guide' to the most common quantities, units, symbols, definitions, formulae and circuit diagram symbols used in the field of electrical and electronic engineering. Some 150 graphical symbols have been selected from British Standards Institution 3939, parts 2–13, 1985.

Those who will find this booklet most valuable will be students and staff in colleges, polytechnics and universities, together with technicians and engineers in industries involved in electrical and electronic engineering.

Thanks are due to Mrs Elaine Woolley for her careful typing of the manuscript and to the publishers for their help in the preparation of this booklet.

J.O. Bird
Highbury College of Technology
Portsmouth

STANDARD ELECTRICAL QUANTITY SYMBOLS AND THEIR UNITS

QUANTITY	QUANTITY SYMBOL	UNIT	UNIT SYMBOL
Admittance	Y	siemen	S
Angular frequency	ω	radians per second	rad/s
Area	A	square metres	m^2
Attenuation coefficient (or constant)	α	neper per metre	Np/m
Capacitance	C	farad	F
Charge	Q	coulomb	C
Charge density	σ	coulomb per square metre	C/m^2
Conductance	G	siemen	S
Current	I	ampere	A
Current density	J	ampere per square metre	A/m^2
Efficiency	η	per-unit or per cent	p.u. or %
Electric field strength	E	volt per metre	V/m
Electric flux	Ψ	coulomb	C
Electric flux density	D	coulomb per square metre	C/m^2
Electromotive force	E	volt	V
Energy	W	joule	J
Field strength, electric	E	volt per metre	V/m
Field strength, magnetic	H	ampere per metre	A/m
Flux, electric	Ψ	coulomb	C
Flux, magnetic	Φ	weber	Wb
Flux density, electric	D	coulomb per square metre	C/m^2

QUANTITY	QUANTITY SYMBOL	UNIT	UNIT SYMBOL
Flux density, magnetic	B	tesla	T
Force	F	newton	N
Frequency	f	hertz	Hz
Frequency, angular	ω	radians per second	rad/s
Frequency, rotational	n	revolutions per second	rev/s
Impedance	Z	ohm	Ω
Inductance, self	L	henry	H
Inductance, mutual	M	henry	H
Length	l	metre	m
Loss angle	δ	radian or degrees	rad or °
Magnetic field strength	H	ampere per metre	A/m
Magnetic flux	Φ	weber	Wb
Magnetic flux density	B	tesla	T
Magnetic flux linkage	Ψ	weber	Wb
Magnetising force	H	ampere per metre	A/m
Magnetomotive force	F_m	ampere	A
Mutual inductance	M	henry	H
Number of phases	m	—	—
Number of pole-pairs	p	—	—
Number of turns (of a winding)	N	—	—
Period, Periodic time	T	second	s
Permeability, absolute	μ	henry per metre	H/m
Permeability of free space	μ_0	henry per metre	H/m
Permeability, relative	μ_r	—	—
Permeance	Λ	weber per ampere or per henry	Wb/A or /H
Permittivity, absolute	ε	farad per metre	F/m
Permittivity of free space	ε_0	farad per metre	F/m
Permittivity, relative	ε_r	—	—
Phase-change coefficient	β	radian per metre	rad/m
Potential, Potential difference	V	volt	V

QUANTITY	QUANTITY SYMBOL	UNIT	UNIT SYMBOL
Power, active	P	watt	W
Power, apparent	S	volt ampere	V A
Power, reactive	Q	volt ampere reactive	var
Propagation coefficient (or constant)	γ	—	—
Quality factor, magnification	Q	—	—
Quantity of electricity	Q	coulomb	C
Reactance	X	ohm	Ω
Reflection coefficient	ρ	—	—
Relative permeability	μ_r	—	—
Relative permittivity	ε_r	—	—
Reluctance	R_m	ampere per weber or per henry	A/Wb or /H
Resistance	R	ohm	Ω
Resistance, temperature coefficient of	α	per degree Celsius or per kelvin	/°C or /K
Resistivity	ρ	ohm metre	Ωm
Slip	s	per unit or per cent	p.u. or %
Standing wave ratio	s	—	—
Susceptance	B	siemen	S
Temperature coefficient of resistance	α	per degree Celsius or per kelvin	/°C or /K
Temperature, thermodynamic	T	kelvin	K
Time	t	second	s
Torque	T	newton metre	N m
Velocity	v	metre per second	m/s
Velocity, angular	ω	radian per second	rad/s
Volume	V	cubic metres	m^3
Wavelength	λ	metre	m

(Note that V/m may also be written as Vm^{-1}; A/m^2 as $A\,m^{-2}$; /K as K^{-1}, and so on.)

3

COMMON PREFIXES

PREFIX	NAME	MEANING MULTIPLY BY
E	exa	10^{18}
P	peta	10^{15}
T	tera	10^{12}
G	giga	10^{9}
M	mega	10^{6}
k	kilo	10^{3}
h	hecto	10^{2}
da	deca	10^{1}
d	deci	10^{-1}
c	centi	10^{-2}
m	milli	10^{-3}
μ	micro	10^{-6}
n	nano	10^{-9}
p	pico	10^{-12}
f	femto	10^{-15}
a	atto	10^{-18}

GREEK ALPHABET

Alpha	A	α
Beta	B	β
Gamma	Γ	γ
Delta	Δ	δ
Epsilon	E	ε
Zeta	Z	ζ
Eta	H	η
Theta	Θ	θ
Iota	I	ι
Kappa	K	$\varkappa$
Lambda	Λ	λ
Mu	M	μ
Nu	N	ν
Xi	Ξ	ξ
Omicron	O	o
Pi	Π	π
Rho	P	ρ
Sigma	Σ	σ
Tau	T	τ
Upsilon	Υ	υ
Phi	Φ	ϕ
Chi	X	χ
Psi	Ψ	ψ
Omega	Ω	ω

SOME BASIC DEFINITIONS

Ampere   The ampere is that constant current which, if maintained in two straight parallel conductors of infinite length, of negligible circular cross-section, and placed 1 metre apart in vaccum, would produce between these conductors a force equal to 2×10^{-7} newton per metre of length.

Atom   An atom is the smallest part of an element which can take part in a chemical reaction and which retains the properties of the element. It contains sub-atomic particles called **electrons, protons** and **neutrons**. Protons and neutrons are contained in the central part of an atom called the **nucleus**. Electrons constitute the outer part of all atoms.

An electron has a mass of 9.11×10^{-31} kilogram and a negative charge of 1.602×10^{-19} coulomb.

A proton has a mass of 1.673×10^{-27} kilogram and a positive charge of 1.602×10^{-19} coulomb (i.e. equal to that of the electron).

An equal number of electrons and protons exist within an atom and it is said to be electrically balanced as the negative and positive charges cancel each other out.

A neutron has a mass of 1.675×10^{-27} kilogram and carries no charge (i.e. it is neutral).

Charge Charge is the quantity of electricity and is measured in coulombs.

Conductance Conductance is the reciprocal of resistance and is measured in siemens.

Conductor A conductor is a material which offers a low resistance to the passage of electric current.

Coulomb The coulomb is the quantity of electricity which flows past a given point in an electric circuit when a current of one ampere is maintained for one second. The coulomb is the unit of electrical charge.

Current Current is the rate of movement of charge.

Electromotive force The electromotive force (emf) is the driving influence which tends to produce on electric current in a closed circuit. It is provided by the source and its unit is the volt.

Energy Energy is the capacity of a system to do work. The unit of energy is the joule J.

Insulator An insulator is a material having a high resistance which prevents current flow.

Joule   One joule is the work done or energy transferred when a force of one newton is exerted through a distance of one metre in the direction of the force. The joule is the unit of work done or energy.

Newton   One newton is the force which, when applied to a mass of one kilogram, gives it an acceleration of one metre per second squared. The newton is the unit of force.

Ohm   The unit of electrical resistance is the ohm, which is the resistance between two points of a conductor when a potential difference of one volt applied between these points, produces in the conductor a current of one ampere (the conductor not being a source of any electromotive force).

Potential   The electric potential at a point is the potential difference between that point and earth, the latter being considered at zero potential.

Potential difference   The potential difference (p.d.) between two points in a circuit is the energy transferred due to the passage of a unit of electrical charge. The unit of potential difference is the volt.

Power   Power is the rate of doing work or transferring energy.

Resistance   The resistance in an electric circuit is its opposition to the flow of electric current.

Resistivity The resistivity of a material is the resistance of a unit cube of the material measured between opposite faces of the cube.

Temperature coefficient of resistance The temperature coefficient of resistance of a material is the ratio of the change of resistance per degree change of temperature to the resistance at 0 °C.

Volt One volt is the difference in potential between two points in a conductor when one joule of energy is required to transfer one coulomb of charge. The volt is the unit of electric potential.

Voltage Voltage is the value of an electromotive force or potential difference expressed in volts.

Watt One watt is one joule per second. The watt is the unit of power.

D.C. CIRCUIT THEORY AND ANALYSIS

Ohm's law states that the current I flowing in a conductor is directly proportional to the applied voltage V and inversely proportional to its resistance R, provided the temperature remains constant.

$$I = \frac{V}{R} \text{ or } V = IR \text{ or } R = \frac{V}{I}$$

For the circuit shown in Fig. 1:

(i)   Voltage $V = V_1 + V_2 + V_3$

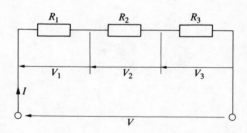

Fig. 1

(ii) Total circuit resistance,

$$R = R_1 + R_2 + R_3$$

(iii) The current I is the same in all parts of the circuit.

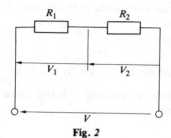

Fig. 2

For the circuit shown in Fig. 2:

$$V_1 = \left(\frac{R_1}{R_1 + R_2}\right) V$$

$$V_2 = \left(\frac{R_2}{R_1 + R_2}\right) V$$

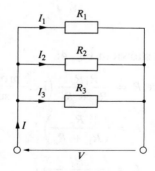

Fig. 3

11

PARALLEL CIRCUIT

For the circuit shown in Fig. 3:

(i)   Current $I = I_1 + I_2 + I_3$

(ii) Total circuit resistance R is determined from

$$\frac{1}{R} = \frac{1}{R_1} + \frac{1}{R_2} + \frac{1}{R_3}$$

i.e. conductance $G = G_1 + G_2 + G_3$

(iii) The potential difference V is the same across each of the resistors.

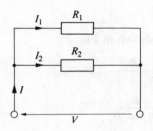

Fig. 4

For the circuit shown in Fig. 4:

Total resistance $R = \dfrac{R_1 R_2}{R_1 + R_2}$ $\left(\text{i.e. } \dfrac{\text{product}}{\text{sum}}\right)$

$$I_1 = \left(\frac{R_2}{R_1 + R_2}\right) I$$

$$I_2 = \left(\frac{R_1}{R_1 + R_2}\right) I$$

POWER AND ENERGY IN D.C. CIRCUITS

Power P in a series or parallel d.c. circuit is given by:

$$P = VI = I^2R = \frac{V^2}{R} \text{ watts}$$

Electrical energy = power × time $= VIt = I^2Rt = \dfrac{V^2t}{R}$

If power is measured in watts and time in seconds, the unit of energy is watt-seconds or **joules**.

If power is measured in kilowatts and time in hours, the unit of energy is **kilowatt-hours** (often called the 'unit of electricity').

$$\textbf{1 kWh} = \textbf{3.6} \times \textbf{10}^6 \textbf{ J}$$

RESISTANCE VARIATION

Resistance $R = \dfrac{\rho l}{a}$ ohms   where ρ = resistivity of the material in ohm metres

l = length of conductor in metres

a = cross-sectional area of conductor in square metres

Resistance at temperature θ °C is given by:

$$R_\theta = R_0(1 + \alpha_0\theta) \quad \text{where } R_0 = \text{resistance at 0 °C}$$

α_0 = temperature coefficient of resistance at 0 °C

$$\frac{R_1}{R_2} = \frac{1 + \alpha_0 \theta_1}{1 + \alpha_0 \theta_2}$$ where R_1 = resistance at
temperature θ_1
R_2 = resistance at
temperature θ_2

TERMINAL P.D. OF A SOURCE

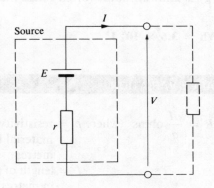

Fig. 5

For the circuit shown in Fig. 5:

terminal p.d., $V = E - Ir$

where E = source emf
r = internal resistance of source
and I = current flowing

14

For d.c. voltage sources connected in series:

total emf = sum of source's emfs
total internal resistance = sum of source's internal resistances

For d.c. voltage sources connected in parallel:

If each source has the same emf and internal resistance,
total emf = emf of one source

total internal resistance of n sources $= \dfrac{1}{n} \times$ internal

resistance of one source

KIRCHHOFF'S LAWS

■ **(i) Current law**

At any junction in an electric circuit, the total current flowing towards that junction is equal to the total current flowing away from the junction, i.e. $\Sigma I = 0$

For the circuit shown in Fig. 6:

$\qquad I_1 + I_2 = I_3 + I_4 + I_5$
$\text{or } \ I_1 + I_2 - I_3 - I_4 - I_5 = 0$

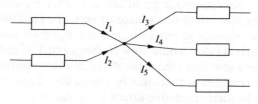

Fig. 6

■ (ii) Voltage law

In any closed loop in a network, the algebraic sum of the voltage drops (i.e. products of current and resistance) taken around the loop is equal to the resultant emf acting in that loop.

For the circuit shown in Fig. 7:

$$E_1 - E_2 = IR_1 + IR_2 + IR_3$$

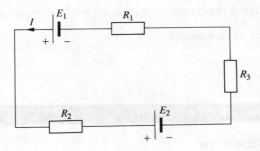

Fig. 7

D.C. CIRCUIT THEOREMS

■ (a) The superposition theorem

In any network made up of linear resistances and containing more than one source of emf, the resultant current flowing in any branch is the algebraic sum of the currents that would flow in that branch if each source was considered separately, all other sources being replaced at that time by resistances equal in value to their respective internal resistances.

■ **(b) Thévenin's theorem**

The current which flows in any branch of a network is the same as that which would flow in that branch if it were connected across a source of electrical energy, the emf of which is equal to the potential difference which would appear across the branch if it were open-circuited, and the internal resistance of which is equal to the resistance which appears across the open-circuited branch terminals when all sources are replaced by resistances equal in value to their internal resistances.

Procedure: To determine the current flowing in a branch containing resistance R of an active network (i.e. one containing a source of emf) using Thévenin's theorem:

(i) remove the resistance R from that branch

(ii) determine the open-circuit voltage E across the break

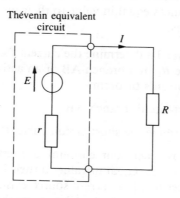

Fig. 8

(iii) remove each source of emf and replace it by a resistance equal in value to its internal resistance and then determine the internal resistance r 'looking-in' at the break

(iv) determine the current I flowing in resistance R from the Thévenin equivalent circuit shown in Fig. 8,

i.e. $I = \dfrac{E}{R + r}$

■ (c) Norton's theorem

The current that flows in any branch of a network is the same as that which would flow in the branch if it were connected across a source of electrical energy, the short-circuit current of which is equal to the current that would flow in a short-circuit across the branch, and the internal resistance of which is equal to the resistance which appears across the open-circuited branch terminals when all sources are replaced by resistances equal in value to their internal resistances.

Procedure: To determine the current flowing in a resistance R_L of a branch AB of an active network using Norton's theorem:

(i) short-circuit branch AB

(ii) determine the short-circuit current I_{SC}

(iii) remove each source of emf and replace them by resistances equal in value to their internal resistances (or, if a current source exists, replace with an open circuit), then determine the resistance R 'looking-in' at a break made between A and B

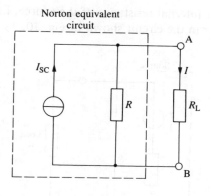

Norton equivalent circuit

Fig. 9

(iv) determine the current I flowing in resistance R_L from the Norton equivalent network shown in Fig. 9

i.e. $I = \left(\dfrac{R}{R + R_L}\right) I_{SC}$

A Thévenin equivalent circuit having emf E and internal resistance r can be replaced by a Norton equivalent circuit containing a current generator I_{SC} and internal resistance R, where:

$$R = r, \ E = I_{SC}R \text{ and } I_{SC} = \frac{E}{r}$$

■ **(d) The maximum power transfer theorem**
The power transferred from a supply source to a load is at its maximum when the resistance of the load is

equal to the internal resistance of the source, i.e. when $R = r$ in the circuit shown in Fig. 10.

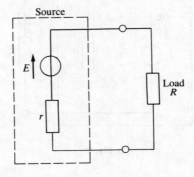

Fig. *10*

CAPACITORS AND CAPACITANCE

An **electric field** is said to exist at a point when a charged body at the point experiences a force by virtue of it being charged.

A **line of electric force** is a line in an electric field, such that the tangent at every point is the direction of the electric force at that point.

A **dielectric** is an insulating medium separating charged surfaces.

Charge $Q = I \times t$ coulombs, where I = current in amperes

and t = time in seconds.

Electric field strength, E, at a point in an electric field, is the force exerted on unit charge placed at that point.

$$E = \frac{V}{d} \text{ volts/metre, where } V = \text{p.d. between two points a distance } d \text{ metres apart.}$$

E is the potential gradient of the field.

Electric flux density D (or charge density σ) is the amount of flux passing through a defined area that is perpendicular to the direction of the flux,

i.e. $D = \sigma = \dfrac{Q}{A}$ coulomb/metre²

Capacitance is a property of two conductors, electrically insulated from each other, which enables them to store energy when a potential difference exists between them.

The unit of capacitance is the **farad** which is defined as the capacitance when the charge stored is one coulomb and the potential difference is one volt.

$$\text{Capacitance } C = \frac{Q}{V} \text{ farads}$$

$$\text{The ratio } \frac{D}{E} = \varepsilon_0 \varepsilon_r = \varepsilon \text{ farads/metre}$$

where ε = absolute permittivity of the dielectric

ε_0 = permittivity of free space = 8.85×10^{-12} F/m

ε_r = relative permittivity

Relative permittivity =

$$\frac{\text{flux density of the field in the dielectric}}{\text{flux density of the field in vacuum (or air)}}$$

Alternatively, relative permittivity is the ratio of the capacitance of a capacitor having a certain material as dielectric to the capacitance of that capacitor with vacuum (or air) as dielectric.

For a **parallel-plate capacitor,**

$$\text{capacitance } C = \frac{\varepsilon_0 \varepsilon_r A (n-1)}{d} \text{ farads,}$$

where A = area of a plate in square metres

d = plate spacing (or dielectric thickness) in metres

n = number of plates

22

☐ *Capacitors connected in parallel and in series*

For n capacitors connected in parallel,

equivalent capacitance, $C_T = C_1 + C_2 + C_3 + \ldots + C_n$

and total charge, $Q_T = Q_1 + Q_2 + Q_3 + \ldots + Q_n$

For n capacitors connected in series, equivalent capacitance C_T is given by:

$$\frac{1}{C_T} = \frac{1}{C_1} + \frac{1}{C_2} + \frac{1}{C_3} + \ldots + \frac{1}{C_n}$$

and the charge is the same on each capacitor.

Dielectric strength E_m is the potential gradient necessary to cause breakdown of an insulating medium,

i.e. $E_m = \dfrac{V_m}{d}$ volts/metre,

where V_m = maximum p.d. between two points a distance d metres apart.

The **energy stored** W in a capacitor is given by:

$$W = \tfrac{1}{2}CV^2 = \tfrac{1}{2}QV \text{ joules}$$

■ **Dielectric loss and loss angle**

☐ *Series representation of losses*

In the circuit diagram of Fig. 11, R_S is the **equivalent** series loss resistance of the capacitor C_S.

From the phasor diagram shown in Fig. 11,

Loss angle of capacitor, $\delta = (90° - \phi)$ where ϕ = angle between applied voltage V and current I.

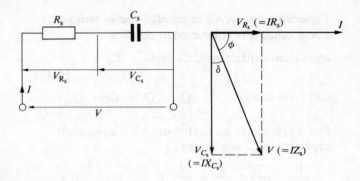

Fig. 11

$$\tan \delta = R_S \omega C_S$$

Power factor of capacitor, $\cos \phi \approx \tan \delta \approx \delta$ (since δ is always small).

□ *Parallel representation of losses*

In the circuit diagram of Fig. 12, R_p is the **equivalent** parallel loss resistance of the capacitor C_p.

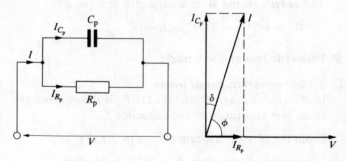

Fig. 12

From the phasor diagram shown in Fig. 12,

$$\tan \delta = \frac{1}{R_p \omega C_p}$$

Power factor of capacitor, **cos ϕ ≈ tan δ ≈ δ** (since δ is always small).

Dielectric power loss = $V^2 \omega C \tan \delta$ watts (where $C = C_S = C_p$).

ELECTROMAGNETISM AND MAGNETIC CIRCUITS

A **magnetic field** is the region in the vicinity of a permanent magnet or a conductor carrying an electric current in which a magnetic pole experiences a mechanical force caused by the magnet or current.

A **permanent magnet** is one in which the material used exhibits magnetism without the need for excitation by a current-carrying coil.

A **line of force** is a line in a magnetic field, such that the tangent at every point is the direction of the magnetic force at that point.

Lines of magnetic force (or flux) are imaginary lines in the magnetic field to aid visualisation of the distribution and density of that field. The direction indicated by the north-seeking end of a compass needle is taken to be the direction of the magnetic field. Each line of force forms a closed path, the lines never intersect and they always have a definite direction.

The magnetic field of a bar magnet can be represented pictorially by the lines of force as shown in Fig. 13; external to the magnet, the direction of the magnetic field is **north to south**.

The **laws of magnetic attraction and repulsion** can be demonstrated by using two bar magnets: with unlike poles

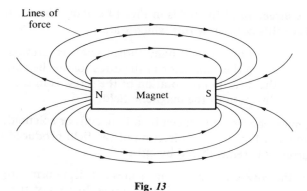

Lines of
force

Fig. *13*

adjacent, attraction occurs; with like poles adjacent,
repulsion occurs.

The magnetic field produced by an **electric current** through
a long, straight conductor forms a circular pattern with
the current-carrying conductor at the centre. The effect is
portrayed in Fig. 14 where the convention adopted is:
current flowing **away** from the viewer is shown as ⊗
current flowing **towards** the viewer is shown as ⊙

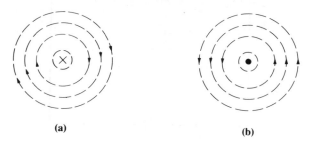

(a)   (b)

Fig. *14*

27

The **direction** of the fields in Fig. 14 is remembered by **Maxwell's corkscrew rule** which states:

> If a normal right-hand thread screw is screwed along the conductor in the direction of the current, the direction of rotation of the screw is in the direction of the magnetic field.

A magnetic field produced by a long coil, or solenoid, is shown in Fig. 15. The direction of the field produced by current I is remembered by either:

(i) the **corkscrew rule**, which states that if a normal right-hand thread screw is placed along the axis of the solenoid and is screwed in the direction of the current it moves in the direction of the magnetic field inside of the solenoid (i.e. points in the direction of the north pole); or

(ii) the **grip rule**, which states that if the coil is gripped

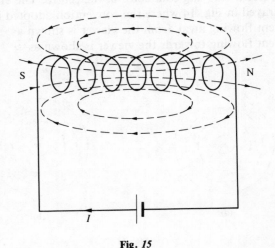

Fig. 15

with the right hand with the fingers pointing in the direction of the current, then the thumb, outstretched parallel to the axis of the solenoid, points in the direction of the magnetic field inside the solenoid (i.e. points in the direction of the north pole).

The **force F between two conductors** d metres apart carrying currents I_1 and I_2 respectively is given by:

$$F = \frac{2 \times 10^{-7} I_1 I_2}{d} \text{ newtons/metre}$$

Magnetic flux density B is the amount of flux passing through a defined area that is perpendicular to the direction of the flux,

i.e. $B = \dfrac{\Phi}{A}$ tesla (where 1 T = 1 Wb/m²)

The **force F on a current-carrying conductor** in a magnetic field when the field is at right angles to the conductor is given by:

$F = BIl$ newtons   where l is the length of conductor perpendicular to the magnetic field, in metres.

When the conductor and the magnetic field are at an angle θ to each other, then:

$F = BIl \sin \theta$ newtons

The direction of the force exerted on a current-carrying conductor in a magnetic field can be predetermined by using **Fleming's left-hand rule** (the motor rule) which states:

Let the thumb, first finger and second finger of

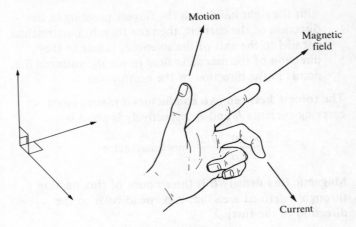

Fig. 16

the left hand be extended such that they are all at
right angles to each other, as shown in Fig. 16. If
the first finger points in the direction of the
magnetic field, the second finger points in the
direction of the current, then the thumb will point
in the direction of the motion of the conductor.

Summarising:
<u>F</u>irst finger   − <u>F</u>ield
Se<u>C</u>ond finger   − <u>C</u>urrent
Thu<u>M</u>b   − <u>M</u>otion

Magnetomotive force (mmf), F_m, is the cause of the
existence of a magnetic flux in a magnetic circuit

$$F_m = NI \text{ amperes}$$ where N = number of
conductors (or
turns)
I = current in
amperes

Magnetic field strength H is the magnetomotive force per unit length of the magnetic circuit,

i.e. $H = \dfrac{NI}{l}$ amperes/metre, where l = mean length of flux path, in metres

Hence $F_m = NI = Hl$ amperes.

For a vacuum and non-magnetic materials:

$$\frac{B}{H} = \mu_0$$

where constant $\mu_0 = 4\pi \times 10^{-7}$ henry/metre, the permeability of free space.

Relative permeability μ_r is the ratio of the flux density produced in a material to the flux density produced in a vacuum (or air) by the same magnetic field strength.

For all media other than free space:

$$\frac{B}{H} = \mu_0\mu_r = \mu \quad \text{where } \mu = \text{absolute permeability}$$

Ferromagnetism is the property of certain materials subjected to a magnetising field which causes induced magnetism which greatly aids the applied field. Such materials are strongly attracted to a magnetic pole and have high relative permeabilities which are greatly dependent on the applied magnetising field. Examples of such materials include iron, steel, nickel and cobalt.

Diamagnetism is a property of certain materials subjected to a magnetising field which causes induced magnetism which opposes the applied field. Such materials are

31

repelled by a magnetic pole and have a relative permeability less than unity.

Paramagnetism is a property of certain materials subjected to a magnetising field which causes induced magnetism which slightly aids the magnetising field. Such materials are attracted very feebly by a magnetic pole and their relative permeability is slightly greater than unity, being almost independent of the magnetising field.

Reluctance R_m is the 'magnetic resistance' of a magnetic circuit to the presence of magnetic flux.

$$R_m = \frac{F_m}{\Phi} \text{ amperes/weber} \quad \text{(Note that 1 A/Wb} = 1 \text{ H}^{-1})$$

Permeance Λ is the magnetic flux per ampere of total magnetomotive force in the path of a magnetic field.

$$\Lambda = \frac{\Phi}{F_m} = \frac{1}{R_m} \text{ weber/ampere} \quad \text{(Note that } 1 \text{ Wb/A} = 1 \text{ H)}$$

Hysteresis is the 'lagging' effect of flux density whenever there are changes in the magnetic field strength. For the **hysteresis loop** shown in Fig. 17:

> OX = remanence (or residual flux density)
> OY = coercive force
> AA' = saturation flux density

☐ *Hysteresis loss*
  Energy is expended in taking a ferromagnetic material through a cycle of magnetisation (i.e. a hysteresis loop); this energy appears as heat in the specimen and is called the hysteresis loss.

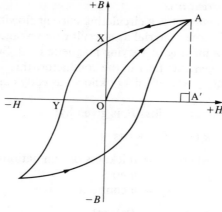

Fig. 17

If a hysteresis loop is plotted to a scale of 1 cm = α A/m along the horizontal axis and 1 cm = β T along the vertical axis, and if A represents the area of the loop in square centimetres, then:

hysteresis loss/cycle = $A\alpha\beta$ joules/metre³

Hysteresis loss, $P_h = k_h v f (B_m)^n$ watts

where v	=	volume of specimen in cubic metres
f	=	frequency in hertz
B_m	=	maximum flux density in teslas
k_h	=	a constant for a given specimen and given range of flux density
n	=	Steinmetz index

33

☐ **Eddy current loss**

An eddy current is a circulating current flowing in a magnetic core material as a result of a voltage induced in it by a moving or varying magnetic field. Such currents generate heat in the conductor, this representing wasted energy known as eddy current loss.

The eddy current loss P_e is given by:

$$P_e = k_e(B_m)^2 f^2 t^2 \text{ watts}$$

$$\text{where } t = \text{thickness of laminations in metres}$$
$$k_e = \text{a constant}$$

Total core loss, $P_c = P_h + P_e$

ELECTROMAGNETIC INDUCTION AND INDUCTANCE

Electromagnetic induction is the production of an electro-motive force when a conductor moves relatively to a magnetic field.

Faraday's laws of electromagnetic induction state:

(i) An induced emf is set up in a circuit whenever the magnetic flux linking with that circuit changes.

(ii) The magnitude of the induced emf in any circuit is proportional to the rate of change of the magnetic flux linking the circuit.

Lenz's law states:

The direction of an induced emf is always such that it tends to set up a current opposing the motion or the change of flux responsible for inducing that emf.

The induced emf e in a coil of N turns is given by:

$$e = -N\frac{d\Phi}{dt} \text{ volts where } \frac{d\Phi}{dt} = \text{the rate of change of flux in webers/second}$$

If the flux linking one turn in a circuit changes by one

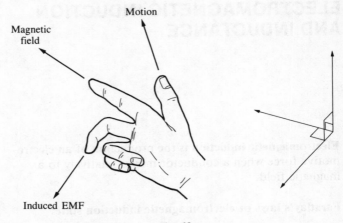

Magnetic field

Motion

Induced EMF

Fig. *18*

weber in one second, a voltage of one volt will be induced in that turn.

Fleming's right-hand rule (generator rule) states:

Let the thumb, first finger and second finger of the right hand be extended such that they are all at right angles to each other, as shown in Fig. 18. If the first finger points in the direction of the magnetic field, the thumb points in the direction of motion of the conductor relative to the magnetic field, then the second finger will point in the direction of the induced emf.

Summarising:

<u>F</u>irst finger   —   <u>F</u>ield
Thu<u>M</u>b   —   <u>M</u>otion
S<u>E</u>cond finger —   <u>E</u>mf

The **induced emf** *E* set up between the ends of a conductor when moved perpendicular to a magnetic field is given by:

$E = Blv$ volts where B = flux density in teslas
l = length of conductor in the magnetic field, in metres
v = conductor velocity in metres/second

When the conductor moves at an angle θ to the magnetic field, then:

$E = Blv \sin \theta$ volts

Flux linkage is the product of the magnitude of magnetic flux and the number of turns of a coil with which the flux is linked.

■ **Inductance**

Inductance is the name given to the property of a coil whereby there is an emf induced into the coil as the result of a change of flux linkages produced by a current change.

(i) When the emf is induced in the same coil as that in which the current is changing, the property is called **self-inductance, L**.

(ii) When the emf is induced in a coil by a change of flux due to current changing in an adjacent coil, the property is called **mutual inductance, M**.

The **induced emf e in a coil of inductance L henrys** is given by:

$$e = -L \frac{di}{dt} \text{ volts} \quad \text{where } \frac{di}{dt} = \text{the rate of change of current in amperes/second}$$

37

The **inductance L of a coil of N turns** is given by:

$$L = \frac{N\Phi}{I} \text{ henry}$$

A circuit has an inductance of one **henry** when an emf of one volt is induced in it by a current changing at the rate of one ampere per second.

The **energy stored W** in the magnetic field of an inductor is given by:

$$W = \tfrac{1}{2}LI^2 \text{ joules}$$

The **mutually induced emf e_2** in a second coil is given by:

$$e_2 = M\frac{\mathrm{d}i_1}{\mathrm{d}t} \text{ volts} \quad \text{where } M = \text{mutual inductance between the two coils, in henrys}$$

$$\frac{\mathrm{d}i_1}{\mathrm{d}t} = \text{the rate of change of current in the first coil in amperes/second.}$$

The mutual inductance between two coils having self-inductance L_1 and L_2 is given by:

$$M = k\sqrt{(L_1 L_2)} \text{ where } k = \text{coupling coefficient}$$

ALTERNATING CURRENTS AND VOLTAGES

A **waveform** is the shape of the curve produced when the magnitude of a quantity that varies (usually in a periodic manner) with time is plotted to a base of time.

Unidirectional waveforms may vary considerably with time but flow in one direction only (i.e. they do not cross the time axis).

Alternating waveforms vary with time, reversing their polarity periodically (i.e. alternately positive and negative).

The process of obtaining unidirectional currents and voltages from alternating currents and voltages is called **rectification.**

Each repetition of a variable quantity, recurring at equal intervals, is called a **cycle**.

The **period or periodic time** T of a waveform is the time taken for an alternating quantity to complete one cycle.

The number of cycles completed in one second is called the **frequency** f of that quantity. The unit of frequency is the **hertz, Hz**.

$$T = \frac{1}{f} \text{ seconds or } f = \frac{1}{T} \text{ hertz}$$

□ *Alternating current values*
    (i) **Instantaneous values** are the values of an alternating quantity at any instant in time. They are

represented by small letters, such as i, v, p, and so on.

(ii) The largest value reached in a half-cycle is called the **peak value** or the **maximum value** or the **amplitude** of the waveform. They are represented by I_m, V_m, and so on.

(iii) A **peak-to-peak value** in an alternating waveform is the difference between the maximum and minimum values in a cycle.

(iv) The **average** or **mean value** of a symmetrical alternating quantity (such as a sine wave) is the average value measured over a half-cycle.

For **any waveform:**

Average or mean value $= \dfrac{\text{area under the curve}}{\text{length of base}}$

Average or mean values are represented by V_{av}, I_{av}, and so on.

For a **sine wave:** average or mean value

$= 0.637 \times$ maximum value

(or $\dfrac{2}{\pi} \times$ maximum value)

(v) The **root mean square (r.m.s.)** or **effective value** of an alternating current is that current which will produce the same heating effect as an equivalent direct current.

R.m.s. values are represented by capital letters, such as V, I, E, and so on.

For the **non-sinusoidal waveform** shown in Fig. 19, the r.m.s. value is given by:

$$I = \sqrt{\left(\frac{i_1^2 + i_2^2 + \ldots + i_n^2}{n}\right)}$$

where *n* is the number of intervals used.

For a **sine wave**, r.m.s. value = 0.707 × maximum value

(or $\frac{1}{\sqrt{2}}$ × maximum value)

Form factor $= \dfrac{\text{r.m.s. value}}{\text{average value}}$ (For a sine wave, form factor = 1.11)

Peak factor $= \dfrac{\text{maximum value}}{\text{r.m.s. value}}$ (For a sine wave, peak factor = 1.41)

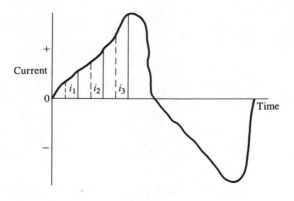

Fig. *19*

Given the general sinusoidal voltage:

$$V = V_m \sin (\omega t \pm \phi) \text{ volts}$$

then (i) Amplitude or maximum value $= V_m$ volts

(ii) Peak to peak value $= 2 V_m$ volts

(iii) Angular frequency $= \omega$ radians/second
(where $\omega = 2\pi f$)

(iv) Frequency $f = \dfrac{\omega}{2\pi}$ hertz

(v) Periodic time $T = \dfrac{1}{f} = \dfrac{2\pi}{\omega}$ seconds

(vi) $\phi =$ angle of lead($+$) or lag($-$),
compared with $v = V_m \sin \omega t$ volts

SINGLE-PHASE A.C. SERIES CIRCUITS

PURE RESISTANCE

In an a.c. circuit containing resistance R only, the current I_R is in phase with the applied voltage V_R, as shown in Fig. 20.

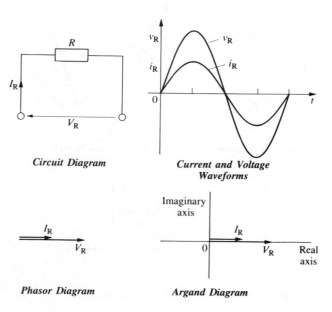

Circuit Diagram

Current and Voltage Waveforms

Phasor Diagram

Argand Diagram

Fig. 20

$$\text{Impedance, } Z = \frac{V_R \angle 0°}{I_R \angle 0°} = R$$

PURE INDUCTANCE

In an a.c. circuit containing pure inductance L only, the current I_L lags the applied voltage V_L by 90° or $\frac{\pi}{2}$ rad, as shown in Fig. 21.

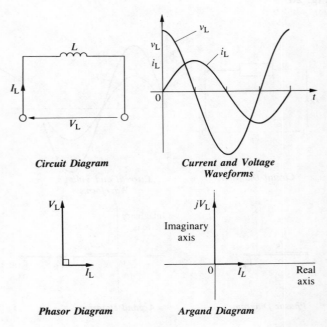

Circuit Diagram

Current and Voltage Waveforms

Phasor Diagram

Argand Diagram

Fig. *21*

Impedance, $Z = \dfrac{V_L \angle 90°}{I_L \angle 0°} = \dfrac{V_L}{I_L} \angle 90°$

$$= X_L \angle 90° \text{ or } jX_L$$

where X_L is the **inductive reactance** given by:

$X_L = 2\pi f L$ ohms   where f = frequency in hertz
L = inductance in henrys

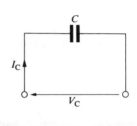

Circuit Diagram

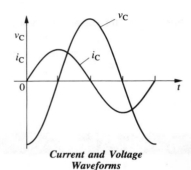

Current and Voltage Waveforms

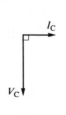

Phasor Diagram

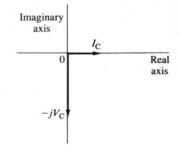

Argand Diagram

Fig. 22

45

PURE CAPACITANCE

In an a.c. circuit containing pure capacitance C only, the current I_C leads the applied voltage V_C by 90° or $\frac{\pi}{2}$ rad, as shown in Fig. 22.

$$\text{Impedance, } Z = \frac{V_C \angle -90°}{I_C \angle 0°} = \frac{V_C}{I_C} \angle -90°$$

$$= X_C \angle -90° \text{ or } -jX_C$$

where X_C is the **capacitive reactance** given by:

$$X_C = \frac{1}{2\pi fC} \text{ ohms} \quad \text{where } C = \text{capacitance in farads}$$

$R-L$ SERIES CIRCUIT

In an a.c. series circuit containing resistance R and inductance L, the applied voltage V is the phasor sum of V_R and V_L and thus the current I lags V by an angle lying between 0° and 90°, shown as angle ϕ in Fig. 23.

From the Argand diagram of Fig. 23 and the voltage triangle shown in Fig. 24:

$$\text{Supply voltage, } V = V_R + jV_L$$

$$|V| = \sqrt{(V_R{}^2 + V_L{}^2)}$$

$$\text{and} \quad \phi = \arctan\left(\frac{V_L}{V_R}\right) \text{ lagging}$$

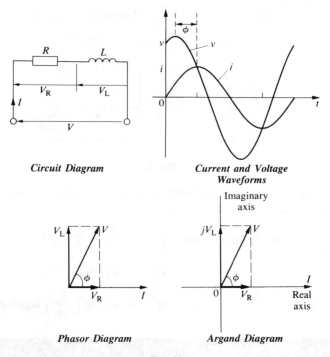

Circuit Diagram

**Current and Voltage
Waveforms**

Phasor Diagram

Argand Diagram

Fig. 23

Voltage Triangle

Fig. 24

47

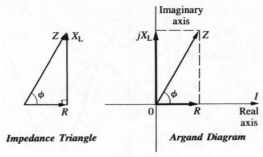

Fig. 25

From the impedance triangle and Argand diagram shown in Fig. 25:

$$\text{Circuit impedance,} \quad Z = \frac{V}{I} = R + jX_\text{L}$$

$$|Z| = \sqrt{(R^2 + X_\text{L}^2)}$$

$$\text{and} \quad \phi = \arctan\left(\frac{X_\text{L}}{R}\right) \text{ lagging}$$

$R - C$ SERIES CIRCUIT

In an a.c. series circuit containing resistance R and capacitance C, the applied voltage V is the phasor sum of V_R and V_C and thus the current I leads V by an angle lying between 0° and 90°, shown as angle α in Fig. 26.

From the Argand diagram of Fig. 26 and the voltage triangle shown in Fig. 27:

$$\text{Supply voltage,} \quad V = V_\text{R} - jV_\text{C}$$

$$|V| = \sqrt{(V_\text{R}^2 + V_\text{C}^2)}$$

$$\text{and} \quad \alpha = \arctan\left(\frac{V_\text{C}}{V_\text{R}}\right) \text{ leading}$$

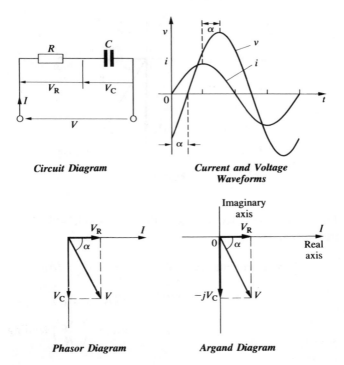

Circuit Diagram

Current and Voltage Waveforms

Phasor Diagram

Argand Diagram

Fig. 26

Voltage Triangle

Fig. 27

49

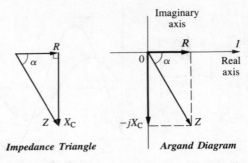

Fig. 28

From the impedance triangle and Argand diagram shown in Fig. 28:

$$\text{Circuit impedance,} \quad Z = \frac{V}{I} = R - jX_C$$

$$|Z| = \sqrt{(R^2 + X_C^2)}$$

$$\text{and} \quad \alpha = \arctan\left(\frac{X_C}{R}\right) \text{ leading}$$

$R-L-C$ SERIES CIRCUIT

In an a.c. circuit containing resistance R, inductance L and capacitance C, as shown in Fig. 29, the applied voltage V is the phasor sum of V_R, V_L and V_C. V_L and V_C are in anti-phase and there are three phasor diagrams possible, each depending on the relative values of V_L and V_C.

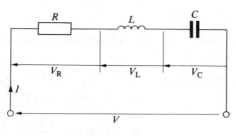

Fig. 29

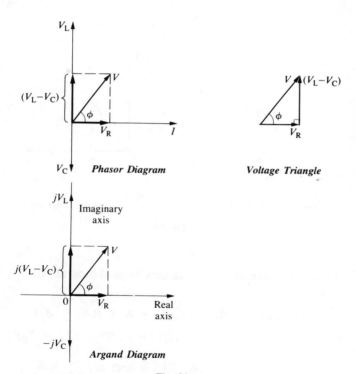

Phasor Diagram

Voltage Triangle

Argand Diagram

Fig. 30

(i) When $V_L > V_C$

From the phasor diagram, voltage triangle and Argand diagram of Fig. 30:

$$\text{Supply voltage,} \quad V = V_R + j(V_L - V_C)$$

$$|V| = \sqrt{[V_R{}^2 + (V_L - V_C)^2]}$$

$$\text{and} \quad \phi = \arctan\left(\frac{V_L - V_C}{V_R}\right) \text{lagging}$$

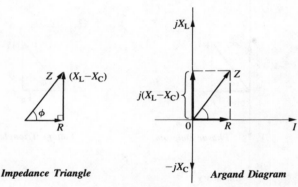

Impedance Triangle   **Argand Diagram**

Fig. 31

From the impedance triangle and Argand diagram of Fig. 31:

$$\text{Circuit impedance,} \quad Z = R + j(X_L - X_C)$$

$$|Z| = \sqrt{[R^2 + (X_L - X_C)^2]}$$

$$\text{and} \quad \phi = \arctan\left(\frac{X_L - X_C}{R}\right) \text{lagging}$$

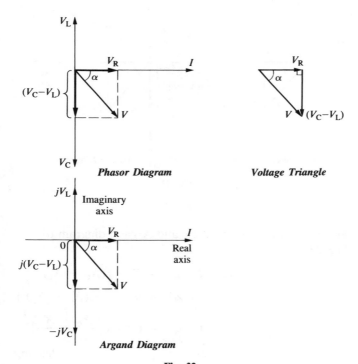

Phasor Diagram

Voltage Triangle

Argand Diagram

Fig. 32

(ii) When $V_C > V_L$

From the phasor diagram, voltage triangle and Argand diagram of Fig. 32:

$$\text{Supply voltage,} \quad V = V_R - j(V_C - V_L)$$

$$|V| = \sqrt{[V_R^2 + (V_C - V_L)^2]}$$

$$\text{and} \quad \alpha = \arctan\left(\frac{V_C - V_L}{V_R}\right) \text{leading}$$

53

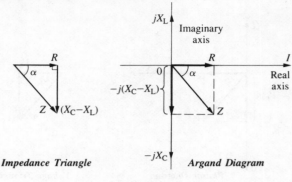

Impedance Triangle   **Argand Diagram**

Fig. 33

From the impedance triangle and Argand diagram of Fig. 33:

$$\text{Circuit impedance,} \quad Z = R - j(X_C - X_L)$$

$$|Z| = \sqrt{[R^2 + (X_C - X_L)^2]}$$

$$\text{and} \quad \alpha = \arctan\left(\frac{X_C - X_L}{R}\right)$$

leading

(iii) **When $V_L = V_C$, series resonance occurs (see page 60)**

GENERAL SERIES CIRCUIT

In an a.c. circuit containing impedances Z_1, Z_2, Z_3, ..., Z_n, connected in series, the total impedance Z_T is given by:

$$Z_T = Z_1 + Z_2 + Z_3 + \ldots + Z_n$$

SINGLE-PHASE A.C. PARALLEL NETWORKS

Admittance Y is the ratio of the current I flowing in an a.c. circuit to the supply voltage V, i.e. it is the reciprocal of impedance Z. The unit of admittance is the **siemen S**.

$$Y = \frac{I}{V} = \frac{1}{Z} \text{ siemens}$$

In complex form, $Y = G + jB$ where G = conductance in siemens

B = susceptance in siemens

Given $Z = R + jX$

$$\text{then } Y = \frac{1}{R + jX} = \frac{R - jX}{R^2 + X^2}$$

$$\text{i.e. } Y = \frac{R}{R^2 + X^2} - \frac{jX}{R^2 + X^2}$$

Hence conductance $G = \dfrac{R}{R^2 + X^2}$

and susceptance $B \quad = \dfrac{-X}{R^2 + X^2}$

■ General parallel network

In an a.c. network containing impedances Z_1, Z_2, Z_3, ... Z_n, connected in parallel, the total equivalent impedance Z_T is determined from:

$$\frac{1}{Z_T} = \frac{1}{Z_1} + \frac{1}{Z_2} + \frac{1}{Z_3} + \ldots + \frac{1}{Z_n}$$

or total admittance, $Y_T = Y_1 + Y_2 + Y_3 + \ldots + Y_n$

For two impedances, Z_1 and Z_2, connected in parallel:

Total impedance $Z_T = \dfrac{Z_1 Z_2}{Z_1 + Z_2} \left(\text{i.e. } \dfrac{\text{product}}{\text{sum}} \right)$

Current flowing in Z_1, $\quad I_1 = \left(\dfrac{Z_2}{Z_1 + Z_2} \right) I$

where I = supply current

Current flowing in Z_2, $\quad I_2 = \left(\dfrac{Z_1}{Z_1 + Z_2} \right) I$

POWER IN A.C. CIRCUITS

The **average power P** in a **purely resistive a.c. circuit** is given by:

$$P = VI = I^2R = \frac{V^2}{R} \text{ watts} \quad \text{where } V \text{ and } I \text{ are r.m.s. values}$$

The average power in a **purely inductive** or a **purely capacitive** a.c. circuit is zero.

The average power P in a circuit containing resistance and inductance and/or capacitance, whether in series or in parallel, is given by:

$$P = VI \cos \phi \text{ watts} \quad \text{where sinusoidal supply voltage } V \text{ and current } I \text{ are r.m.s. values and } \phi \text{ is the phase angle between } V \text{ and } I$$

or $\quad P = I_R^2R \text{ watts} \quad$ where I_R is the r.m.s. current flowing in resistance R

□ *Power triangle*
A phasor diagram for an inductive circuit is shown in Fig. 34(a). The horizontal component of voltage V is $V \cos \phi$ and the vertical component of V is $V \sin \phi$.

If each of the voltage phasors of triangle oab is multiplied by current I, Fig. 34(b) is produced and is known as the power triangle. (The power triangle for a capacitive circuit is shown in Fig. 34(c).)

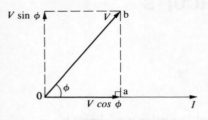

(a) *Phasor Diagram for Inductive Circuit*

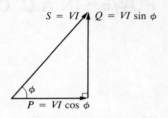

(b) *Power Triangle for Inductive Circuit*

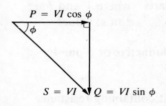

(c) *Power Triangle for Capacitive Circuit*

Fig. *34*

Each side of the power triangle represents a particular type of power:

True or active power $P = VI \cos \phi$ watts (W)

Apparent power $S = VI$ voltamperes (V A)

Reactive power $Q = VI \sin \phi$ vars (var)

$$\text{Power factor} = \frac{\text{true power } P}{\text{apparent power } S}$$

For sinusoidal voltages and currents:

$$\text{Power factor} = \cos \phi = \frac{R}{Z} \text{ (from the impedance triangle)}$$

■ Use of complex numbers for determination of power

If voltage $V = (a + jb)$ volts and
current $I = (c + jd)$ amperes, then:

$$\text{power } P = ac + bd$$
$$\text{reactive power } Q = bc - ad$$
$$\text{and apparent power } S = VI^*$$

where I^* is the conjugate of I, i.e. $(c - jd)$

$$= (a + jb)(c - jd)$$
$$= (ac + bd) + j(bc - ad)$$

i.e.
$$S = P + jQ$$

SERIES AND PARALLEL RESONANCE AND Q-FACTOR

When the voltage applied to an electrical network containing resistance, inductance and capacitance is in phase with the resulting current, the circuit is said to be **resonant**.

SERIES RESONANCE

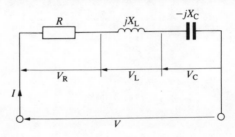

Fig. 35

For the a.c. series circuit shown in Fig. 35:

Impedance $Z = R + j(X_L - X_C)$

and resonance occurs when $(X_L - X_C) = 0$

from which, resonant frequency, $f_r = \dfrac{1}{2\pi\sqrt{(LC)}}$ hertz

At resonance, $V_L = V_C$

$$Z = R, \text{ the minimum circuit impedance possible}$$

and $I = \dfrac{V}{R}$, the maximum current possible, in an $L-C-R$ series circuit

Q-factor is a figure of merit for a resonant device, such as an $L-R-C$ circuit. Q-factor is an abbreviation for **quality factor** and indicates the extent to which the inductor or capacitor approximates to a pure reactance.

For a series a.c. circuit at resonance:

$$Q = \frac{\omega_r L}{R} = \frac{1}{\omega_r CR} = \frac{1}{R}\sqrt{\left(\frac{L}{C}\right)}$$

Q is also the voltage magnification factor in a series circuit,

i.e. $Q = \dfrac{V_C \text{ (or } V_L)}{V}$

BANDWIDTH

Figure 36 shows how current varies with frequency in an $L-C-R$ series circuit. A and B are the points at which current is 0.707 of its maximum value (which occurs at resonance). The points, corresponding to frequencies f_1 and f_2, are called the **half-power points**, and $(f_2 - f_1)$ is called the **bandwidth**. The half-power points are also referred to as the '**-3 dB points**'.

At the half-power frequencies, impedance $Z = \sqrt{2}R$,

61

$$f_1 = \frac{-R + \sqrt{\left(R^2 + \dfrac{4L}{C}\right)}}{4\pi L} \text{ and}$$

$$f_2 = \frac{R + \sqrt{\left(R^2 + \dfrac{4L}{C}\right)}}{4\pi L}$$

Also, resonant frequency,

$$f_r = \sqrt{(f_1 f_2)} \text{ and } Q = \frac{f_r}{(f_2 - f_1)} = \frac{f_r}{\text{bandwidth}}$$

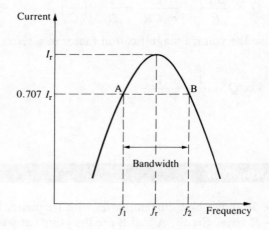

Fig. *36*

Selectivity is the ability of a circuit to respond more readily to signals of a particular frequency to which it is tuned than to signals of other frequencies.

PARALLEL RESONANCE

For the 'ideal' a.c. parallel network shown in Fig. 37:

Total network admittance, $Y = \dfrac{1}{R} + \dfrac{1}{jX_L} + \dfrac{1}{-jX_C}$

$$= \dfrac{1}{R} - \dfrac{j}{\omega L} + j\omega C$$

i.e. $Y = \dfrac{1}{R} + j\left(\omega C - \dfrac{1}{\omega L}\right)$

At resonance, $\omega C - \dfrac{1}{\omega L} = 0$, from which, resonant

frequency, $f_r = \dfrac{1}{2\pi\sqrt{(LC)}}$ hertz

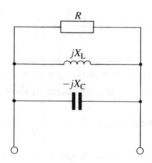

Fig. 37

For any parallel network at resonance:

(i) total network admittance Y is a minimum
(ii) total network impedance Z is a maximum
(iii) supply current I is a minimum, and

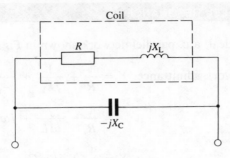

Fig. 38

(iv) an expression for the resonant frequency f_r may be obtained by making the imaginary part of the complex expression for admittance equal to zero and then transposing for f_r

For the parallel a.c. network shown in Fig. 38:

$$\text{Resonant frequency } f_r = \frac{1}{2\pi} \sqrt{\left(\frac{1}{LC} - \frac{R^2}{L^2}\right)} \text{ hertz}$$

Since the current at resonance is in phase with the voltage, the impedance of the network acts as a resistance; this resistance is known as the **dynamic resistance R_D.**

$$\text{Dynamic resistance } R_D = \frac{L}{CR} \text{ ohms}$$

For the parallel a.c. network shown in Fig. 39:

$$\text{Resonant frequency } f_r = \frac{1}{2\pi\sqrt{(LC)}} \sqrt{\left(\frac{R_L{}^2 - \dfrac{L}{C}}{R_C{}^2 - \dfrac{L}{C}}\right)} \text{ hertz}$$

The Q-factor of a parallel resonant network is the ratio of the current circulating in the parallel branches of the

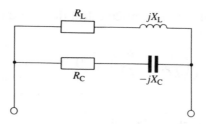

Fig. 39

network to the supply current, i.e. Q-factor is a measure of the **current magnification.**

For the parallel network shown in Fig. 40:

$$Q = \frac{I_C}{I_r} = \frac{\omega_r L}{R}$$

Around the closed loop comprising the coil and the capacitor of Fig. 40, the energy would naturally resonate at a frequency given by that for a series $L-C-R$ circuit, i.e. $\frac{1}{2\pi\sqrt{(LC)}}$. This frequency is termed the **natural frequency f_n,** and the frequency of resonance seen at the

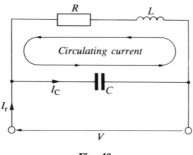

Fig. 40

terminals of Fig. 40 is often called the **forced resonant frequency**, f_r, i.e.

$$\frac{1}{2\pi}\sqrt{\left(\frac{1}{LC} - \frac{R^2}{L^2}\right)}$$

(For a series a.c. circuit, the forced and natural frequencies coincide.)

For the network of Fig. 40:

$$f_r = f_n \sqrt{\left(1 - \frac{1}{Q^2}\right)} \text{ hertz}$$

For any parallel a.c. network:

$$Q = \frac{f_r}{(f_2 - f_1)} \text{ and } f_r = \sqrt{(f_1 f_2)} \text{ (as for a series circuit)}$$

where f_1 and f_2 are the half-power frequencies.

A.C. NETWORK ANALYSIS

The laws which determine the currents and voltage drops in a.c. networks are:

(a) current $I = \dfrac{V}{Z}$

(b) the laws for impedances in series and parallel, i.e.

total impedance, $Z_T = Z_1 + Z_2 + \ldots + Z_n$
for n impedances connected in series

and $\dfrac{1}{Z_T} = \dfrac{1}{Z_1} + \dfrac{1}{Z_2} + \ldots + \dfrac{1}{Z_n}$
for n impedances connected in parallel

(or $Y_T = Y_1 + Y_2 + \ldots + Y_n$)

and (c) **Kirchhoff's laws**, which may be stated as:

(i) At any junction in an electrical circuit the phasor sum of the currents flowing towards that junction is equal to the phasor sum of the currents flowing away from the junction.

(ii) In any closed loop in a network, the phasor sum of the voltage drops (i.e. the products of current and impedance) taken around the loop is equal to the phasor sum of the emfs acting in that loop.

Determinants may be used to solve simultaneous equations resulting from the application of Kirchhoff's laws or when using mesh or nodal analysis.

Two unknowns. When solving linear simultaneous equations in two unknowns using determinants:

(i) the equations are initially written in the form

$$a_1x + b_1y + c_1 = 0$$
$$a_2x + b_2y + c_2 = 0$$

and (ii) the solution is given by

$$\frac{x}{D_x} = -\frac{y}{D_y} = \frac{1}{D}$$

where $D_x = \begin{vmatrix} b_1 & c_1 \\ b_2 & c_2 \end{vmatrix}$ i.e. the determinant of the coefficients left when the x-column is 'covered-up',

$D_y = \begin{vmatrix} a_1 & c_1 \\ a_2 & c_2 \end{vmatrix}$ i.e. the determinant of the coefficients left when the y-column is 'covered-up',

and $D = \begin{vmatrix} a_1 & b_1 \\ a_2 & b_2 \end{vmatrix}$ i.e. the determinant of the coefficients left when the constants-column is 'covered-up'.

A '2 by 2' determinant $\begin{vmatrix} a & b \\ c & d \end{vmatrix}$ is evaluated as $ad - bc$

Three unknowns. When solving linear simultaneous equations in three unknowns using determinants:

(i) the equations are initially written in the form

$$a_1x + b_1y + c_1z + d_1 = 0$$
$$a_2x + b_2y + c_2z + d_2 = 0$$
$$a_3x + b_3y + c_3z + d_3 = 0$$

and   (ii) the solution is given by:

$$\frac{x}{D_x} = -\frac{y}{D_y} = \frac{z}{D_z} = -\frac{1}{D}$$

where $D_x = \begin{vmatrix} b_1 & c_1 & d_1 \\ b_2 & c_2 & d_2 \\ b_3 & c_3 & d_3 \end{vmatrix}$ i.e. the determinants of the coefficients left when the x-column is 'covered-up'.

$$D_y = \begin{vmatrix} a_1 & c_1 & d_1 \\ a_2 & c_2 & d_2 \\ a_3 & c_3 & d_3 \end{vmatrix} \quad D_z = \begin{vmatrix} a_1 & b_1 & d_1 \\ a_2 & b_2 & d_2 \\ a_3 & b_3 & d_3 \end{vmatrix} \quad \text{and} \quad D = \begin{vmatrix} a_1 & b_1 & c_1 \\ a_2 & b_2 & c_2 \\ a_3 & b_3 & c_3 \end{vmatrix}$$

A '3 by 3' determinant $\begin{vmatrix} a & b & c \\ d & e & f \\ g & h & j \end{vmatrix}$ is evaluated as

$$a \begin{vmatrix} e & f \\ h & j \end{vmatrix} - b \begin{vmatrix} d & f \\ g & j \end{vmatrix} + c \begin{vmatrix} d & e \\ g & h \end{vmatrix} \text{ using the top row.}$$

■ **Application of Kirchhoff's laws**

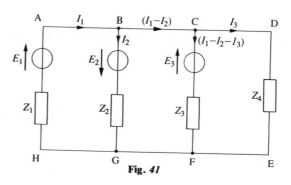

Fig. 41

Applying Kirchhoff's laws to the network of Fig. 41 gives:

69

From loop ABGH and moving clockwise:

$$I_1 Z_1 + I_2 Z_2 = E_1 + E_2 \qquad (1)$$

From loop BCFG and moving anticlockwise:

$$I_2 Z_2 - (I_1 - I_2 - I_3)Z_3 = E_2 + E_3 \qquad (2)$$

From loop CDEF and moving clockwise:

$$-(I_1 - I_2 - I_3)Z_3 + I_3 Z_4 = E_3 \qquad (3)$$

Given values of impedances Z_1 to Z_4 and source emfs E_1 to E_3, then equations (1) to (3) may be used to solve for currents I_1, I_2 and I_3 using determinants.

■ **Application of mesh-current analysis (or Maxwell's theorem)**

Applying mesh-current analysis to the network of Fig. 42 gives:

$$I_1(Z_1 + Z_2) - I_2 Z_2 = E_1 \qquad (4)$$

$$-I_1 Z_2 + I_2(Z_2 + Z_3 + Z_4) - I_3 Z_4 = 0 \qquad (5)$$

$$-I_2 Z_4 + I_3(Z_4 + Z_5) = -E_2 \qquad (6)$$

Given values of impedances Z_1 to Z_5 and source emfs

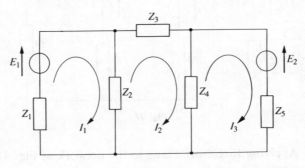

Fig. 42

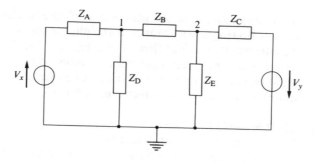

Fig. 43

E_1 and E_2, then equations (4) to (6) may be solved for mesh currents I_1, I_2 and I_3 using determinants.

■ Application of nodal analysis

Applying nodal analysis to the network of Fig. 43 gives:

$$\text{At node 1: } \frac{V_1 - V_x}{Z_A} + \frac{V_1}{Z_D} + \frac{V_1 - V_2}{Z_B} = 0 \qquad (7)$$

$$\text{At node 2: } \frac{V_2 - V_1}{Z_B} + \frac{V_2}{Z_E} + \frac{V_2 + V_y}{Z_C} = 0 \qquad (8)$$

Given values of impedances Z_A to Z_E and source emfs V_x and V_y, then equations (7) and (8) may be solved for node voltages V_1 and V_2 by using determinants.

A.C. CIRCUIT THEOREMS

■ (a) Superposition theorem

In any network made up of linear impedances and

71

containing more than one source of emf, the resultant current flowing in any branch is the phasor sum of the currents that would flow in that branch if each source was considered separately, all other sources being replaced at that time by impedances equal in value to their respective internal impedances.

■ **(b) Thévenin's theorem**

The current which flows in any branch of a circuit is the same as that which would flow in that branch if it were connected across a source of electrical energy, the emf of which is equal to the potential difference which would appear across the branch if it were open-circuited, and the internal impedance of which is equal to the impedance which appears across the open-circuited branch terminals when all sources are replaced by impedances equal in value to their internal impedances.

Procedure: To determine the current flowing in a branch containing impedance Z_L of an active network using Thévenin's theorem:

(i)   remove the impedance Z_L from that branch

(ii) determine the open-circuit voltage E across the break

(iii) remove each source of emf and replace it by an impedance equal in value to its internal impedance and then determine the internal impedance z 'looking-in' at the break

(iv) determine the current i_L flowing in impedance Z_L from the Thévenin equivalent circuit shown in Fig. 44,

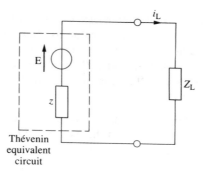

Fig. 44

i.e. $i_L = \dfrac{E}{Z_L + z}$

■ **(c) Norton's theorem**

The current that flows in any branch of a network is the same as that which would flow in that branch if it were connected across a source of electrical energy, the short-circuit current of which is equal to the current that would flow in a short-circuit across the branch, and the internal impedance of which is equal to the impedance which appears across the open-circuited branch terminals when all sources are replaced by impedances equal in value to their internal impedances.

Procedure: To determine the current flowing in an impedance Z_L of a branch AB of an active network using Norton's theorem:

(i)   short-circuit branch AB

(ii) determine the short-circuit current I_{sc}

(iii) remove each source of emf and replace them by impedances equal in value to their internal impedances (or, if a current source exists, replace with an open-circuit), then determine the impedance z 'looking-in' at a break made between A and B

(iv) determine the value of current i_L flowing in impedance Z_L from the Norton equivalent network shown in Fig. 45,

i.e. $i_L = \left(\dfrac{z}{Z_L + z} \right) I_{SC}$

☐ **Thévenin and Norton equivalent circuits**
The Thévenin and Norton networks shown in Fig. 46 are equivalent to each other.

$$I_{SC} = \frac{E}{z} \text{ or } E = I_{SC} z$$

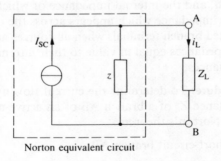

Norton equivalent circuit

Fig. 45

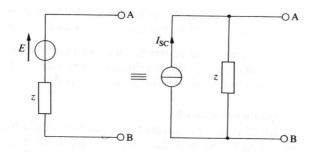

Fig. 46

■ (d) Maximum power transfer theorems
With reference to Fig. 47:

(i) When the load is purely resistive (i.e. $X = 0$) and adjustable, maximum power transfer is achieved when $R = |z| = \sqrt{(r^2 + x^2)}$.

(ii) When both the load and the source impedance are purely resistive (i.e. $X = x = 0$), maximum power transfer is achieved when $R = r$.

(iii) When the load resistance R and reactance X are

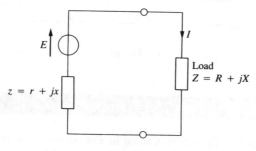

Fig. 47

both independently adjustable, maximum power transfer is achieved when $X = -x$ and $R = r$.

(iv) When the load resistance R is adjustable with reactance X fixed, maximum power transfer is achieved when $R = \sqrt{[r^2 + (x + X)^2]}$.

■ **Impedance matching**

Figure 48 represents a transformer of turns ratio $N_1{:}N_2$ supplying a load impedance Z_L. The input impedance z of the transformer is given by:

$$| z | = \left(\frac{N_1}{N_2}\right)^2 | Z_L |$$

$$\left(\text{assuming that } \frac{V_1}{V_2} = \frac{N_1}{N_2} = \frac{I_2}{I_1}\right)$$

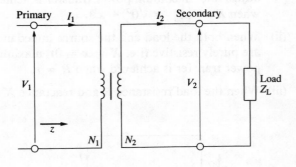

Fig. 48

DELTA-STAR (OR π – T) TRANSFORMATION

The star (or T) section shown in Fig. 49(b) is equivalent to the delta (or π) section shown in Fig. 49(a) when:

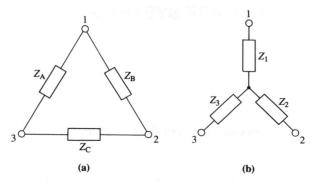

Fig. 49

$$Z_1 = \frac{Z_A Z_B}{Z_A + Z_B + Z_C}$$

$$Z_2 = \frac{Z_B Z_C}{Z_A + Z_B + Z_C}$$

$$Z_3 = \frac{Z_A Z_C}{Z_A + Z_B + Z_C}$$

STAR-DELTA (OR T – π) TRANSFORMATION

The delta (or π) section shown in Fig. 49(a) is equivalent to the star (or T) section shown in Fig. 49(b) when:

$$Z_A = \frac{Z_1 Z_2 + Z_2 Z_3 + Z_3 Z_1}{Z_2}$$

$$Z_B = \frac{Z_1 Z_2 + Z_2 Z_3 + Z_3 Z_1}{Z_3}$$

$$Z_C = \frac{Z_1 Z_2 + Z_2 Z_3 + Z_3 Z_1}{Z_1}$$

THREE-PHASE SYSTEMS

For the **star-connected load** shown in Fig. 50:

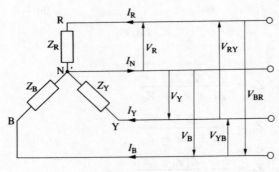

Fig. 50

$$I_L = I_p \quad \text{(where suffix L denotes line and suffix p denotes phase)}$$

For a balanced system:

$$I_R = I_Y = I_B, \ V_R = V_Y = V_B, \ V_{RY} = V_{YB} = V_{BR},$$
$$Z_R = Z_Y = Z_B \text{ and } I_N = 0$$

For a balanced star connection:

$$V_L = \sqrt{3} V_p$$

For the **delta (or mesh)-connected load** shown in Fig. 51:

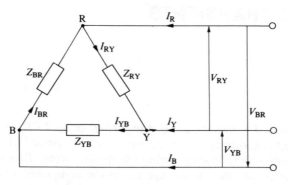

Fig. 51

$$V_{\mathrm{L}} = V_{\mathrm{p}}$$

For a balanced delta connection:

$$I_{\mathrm{L}} = \sqrt{3}I_{\mathrm{p}}$$

■ Power dissipated in a three-phase load

For either a star- or a delta-balanced connection the total power P dissipated is given by:

$$P = \sqrt{3}\,V_{\mathrm{L}}I_{\mathrm{L}} \cos \phi \text{ watts}$$

or $\quad P = 3I_{\mathrm{p}}^2 R_{\mathrm{p}} \text{ watts}$

With the **two-wattmeter method** for measuring power with balanced or unbalanced loads,

Total power = sum of wattmeter readings
= $P_1 + P_2$

The power factor may be determined from:

$$\tan \phi = \sqrt{3}\,\frac{(P_1 - P_2)}{(P_1 + P_2)}$$

79

D.C. TRANSIENTS

When a d.c. voltage applied to a $C-R$ or an $L-R$ series circuit is suddenly changed, there is a brief period of time immediately following during which the current flowing in the circuit and the voltage across the components are changing. These changing values are called **transients**.

SERIES $C-R$ CIRCUIT

■ **(a) Charging**
Assuming no initial charge on capacitance C in Fig. 52 when switch S is closed, then

(i) the capacitor voltage v_C rises exponentially as shown in Fig. 53(a)

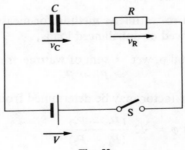

Fig. 52

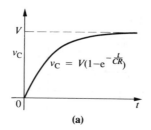

(a)

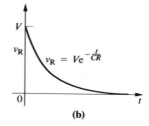

(b)

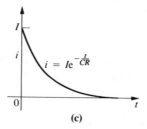

(c)

Fig. 53

(ii)   the resistor voltage v_R decays exponentially as shown in Fig. 53(b) (note $V = v_C + v_R$)

(iii) the current i decays exponentially as shown in Fig. 53(c)

The **time constant** of a circuit is the time for a transient to reach its final state if the initial rate of change were to be maintained.

In a $C-R$ series circuit, the time constant τ is given by:

   $\tau = CR$ **seconds**

In a time equal to τ seconds, v_C rises to 63% of its steady-state value V and i falls to 37% of its initial value I.

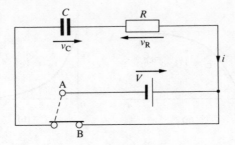

Fig. 54

■ **(b) Discharging**
When the capacitor shown in Fig. 54 is charged with
the switch in position A, and the switch is then moved
to position B, then:

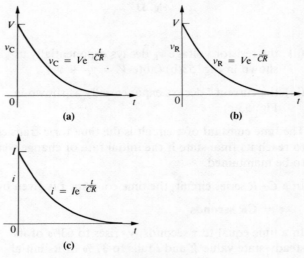

Fig. 55

(i) the capacitor voltage v_C decays exponentially as shown in Fig. 55(a)

(ii) the resistor voltage v_R decays exponentially as shown in Fig. 55(b)

(iii) the current i decays exponentially as shown in Fig. 55(c)

SERIES $L - R$ CIRCUIT

■ (a) Current growth

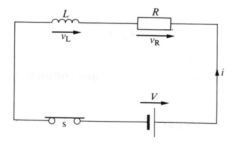

Fig. 56

When switch S in Fig. 56 is closed as shown, then:

(i) the inductor voltage v_L decays exponentially as shown in Fig. 57(a)

(ii) the resistor voltage v_R rises exponentially as shown in Fig. 57(b) (note $V = v_L + v_R$)

(iii) the current i rises exponentially as shown in Fig. 57(c)

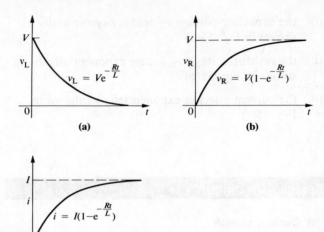

Fig. 57

In an $L-R$ series circuit, the time constant τ is given by:

$$\tau = \frac{L}{R} \text{ seconds}$$

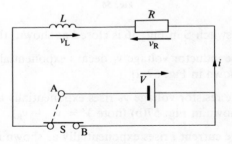

Fig. 58

■ (b) Current decay

When switch S in Fig. 58, initially connected to position A, is moved to position B, then:

(i) the inductor voltage v_L decays exponentially as shown in Fig. 59(a)

(ii) the resistor voltage v_R decays exponentially as shown in Fig. 59(b)

(iii) the current i decays exponentially as shown in Fig. 59(c)

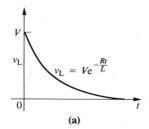

(a)

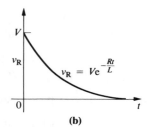

(b)

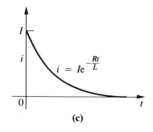

(c)

Fig. 59

SINGLE-PHASE TRANSFORMERS

For the transformer shown in Fig. 60, assuming no losses:

$$\frac{V_1}{V_2} = \frac{N_1}{N_2} = \frac{I_2}{I_1}$$

Transformer efficiency, η, is given by:

$$\eta = \frac{\text{output power}}{\text{input power}} = \frac{\text{output power}}{\text{output power + losses}}$$

$$= \frac{[\, V_2 I_2 \times \text{p.f.}\,]}{[\, V_2 I_2 \times \text{p.f.}\,] + (P_c + I_1^2 R_1 + I_2^2 R_2)}$$

where   p.f. = power factor

P_c = total iron loss in core

$I_1^2 R_1 + I_2^2 R_2$ = total copper loss in primary and secondary windings

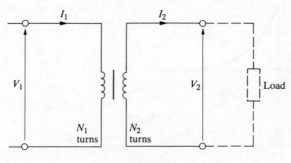

Fig. 60

or $\quad \eta = 1 - \dfrac{\text{losses}}{\text{input power}}$

The condition for **maximum efficiency** is that: copper loss = iron loss.

For any load equal to $n \times$ full load:

$$\eta = \frac{[\,n \times \text{full load VA} \times \text{p.f.}\,]}{[\,n \times \text{full load VA} \times \text{p.f.}\,] + P_{\text{oc}} + n^2 P_{\text{sc}}}$$

where P_{oc} = input power on the open-circuit test (= iron loss)

$\quad\;\; P_{\text{sc}}$ = input power on the short-circuit test with full-load currents (= total copper loss on full load)

The **emf equation of a transformer** is given by:

rms value of primary-induced emf, $E_1 = 4.44 N_1 f \Phi_{\text{m}}$ volts

where Φ_{m} = maximum value of flux in webers

$\quad\; f$ = frequency in hertz

rms value of secondary-induced emf, $E_2 = 4.44 N_2 f \Phi_{\text{m}}$ volts

Transformer no-load current I_0 is given by:

$I_0 = \sqrt{(I_{\text{M}}^2 + I_{\text{c}}^2)}$ where I_{c} = core loss component of current

I_{M} = magnetising component of current

Power factor on no-load $= \cos \phi_0 = \dfrac{I_c}{I_0}$

where ϕ_0 = phase angle between V_1 and I_0.

Total core loss (i.e. iron losses) $= V_1 I_0 \cos \phi_0$ watts.

■ **Equivalent circuit of a transformer**

The equivalent resistance R_e referred to the primary is given by:

$$R_e = R_1 + R_2 \left(\frac{V_1}{V_2}\right)^2 \text{ ohms}$$

where R_1 = primary resistance

R_2 = secondary resistance

The equivalent reactance X_e referred to the primary is given by:

$$X_e = X_1 + X_2 \left(\frac{V_1}{V_2}\right)^2 \text{ ohms}$$

where X_1 = primary inductive reactance

X_2 = secondary inductive reactance

The equivalent impedance Z_e referred to the primary is given by:

$$Z_e = \sqrt{(R_e^2 + X_e^2)} \text{ ohms}$$

The **voltage regulation of a transformer** is the variation of the secondary voltage between no-load and full-load expressed as either a per-unit or a percentage of the no-load voltage, the primary voltage being assumed constant, i.e. per-unit voltage regulation $=$

$$\left(\frac{\text{no-load voltage } - \text{ full-load voltage}}{\text{no-load voltage}} \right)$$

For a load having a power factor of $\cos \phi_2$ lagging:

$$\text{per-unit voltage regulation } = \frac{I_1 Z_e \cos(\phi_e - \phi_2)}{V_1}$$

$$= \frac{I_1 (R_e \cos \phi_2 + X_e \sin \phi_2)}{V_1}$$

or, per-unit voltage regulation $= \dfrac{V_{SC} \cos(\phi_e - \phi_2)}{V_1}$

where $V_{SC} =$ primary voltage on a short-circuit test when full-load currents are flowing in the primary and secondary windings

and $\phi_e =$ phase angle between I_1 and the volt drop due to Z_e

D.C. MACHINES

The **percentage efficiency** η of an electrical machine is given by:

$$\eta = \left(\frac{\text{output power}}{\text{input power}} \right) \times 100\%$$

For a d.c. machine the **generated emf** E is given by:

$$E = \frac{2p\varPhi nZ}{c} \quad \text{where } p = \text{number of pairs of poles}$$

$\varPhi$ = flux per pole, in webers
n = speed in revolutions/second
Z = number of armature conductors
c = $2p$ for a lap-wound machine
= 2 for a wave-wound machine

The **armature mmf at each brush axis** is given by:

$$\text{ampere-turns/pole} = \frac{1}{2} \frac{I}{c} \cdot \frac{Z}{2p}$$

where I = total armature current, in amperes

demagnetising ampere-turns/pole $= \dfrac{1}{2} \dfrac{I}{c} \cdot \dfrac{Z}{2p} \cdot \dfrac{4\theta}{360}$

where θ = brush shift in degrees

distorting ampere-turns/pole $= \dfrac{1}{2} \cdot \dfrac{I}{c} \cdot \dfrac{Z}{2p}\left(1 - \dfrac{4\theta}{360}\right)$

ampere-turns/pole for compensating winding

$$= \frac{1}{2} \cdot \frac{I}{c} \cdot \frac{Z}{2p}\left(\frac{\text{pole arc}}{\text{pole pitch}}\right)$$

For a given d.c. machine, $E \propto \Phi n$

or $E \propto \Phi \omega$

where $\omega = 2\pi n = $ angular velocity
in radians/second

The **power on the shaft of a d.c. machine** is given by:

shaft power $= T\omega$ watts
where $T = $ torque in newton metres

■ Shunt motor

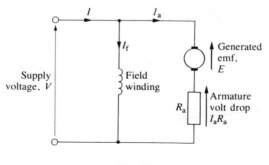

Fig. *61*

For the shunt motor shown in Fig. 61:

Supply voltage $V = E + I_a R_a$

or generated emf $E = V - I_a R_a$

Supply current $I = I_a + I_f$

Torque $T \propto \Phi I_a$

Speed $n \propto \dfrac{E}{\Phi} \propto \dfrac{V - I_a R_a}{\Phi}$

Efficiency of shunt motor,

$$\eta = \left(\frac{VI - I_a^2 R_a - VI_f - C}{VI} \right) \times 100\%$$

where C = sum of iron, friction and windage losses

The **efficiency of a motor is a maximum** when the load is such that:

$$I_a^2 R_a = VI_f + C$$

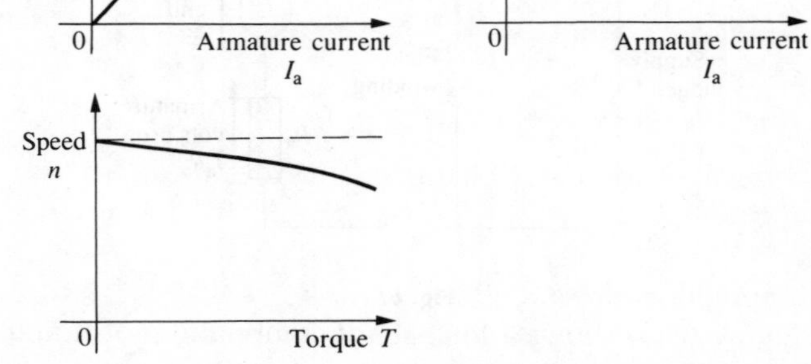

Fig. 62

Typical shunt motor characteristics are shown in Fig. 62.

■ Series motor

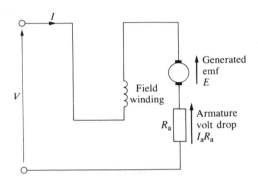

Fig. *63*

For the series motor shown in Fig. 63:

Supply voltage $V = E + IR_a$

or generated emf $E = V - IR_a$

Speed $n\ \alpha\ \dfrac{E}{\Phi}\ \alpha\ \dfrac{V - IR_a}{\Phi}\ \alpha\ \dfrac{V - IR}{I}\ \alpha\ \dfrac{1}{I}$

where $R = R_a + R_f$

Typical series motor characteristics are shown in Fig. 64.

■ Compound motor

Figure 65(a) shows a **long-shunt** compound motor and Fig. 65(b) a **short-shunt** compound motor.

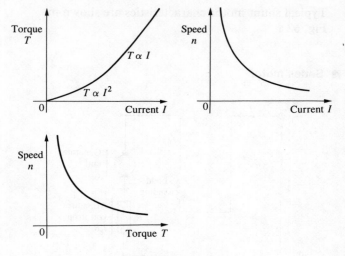

Fig. 64

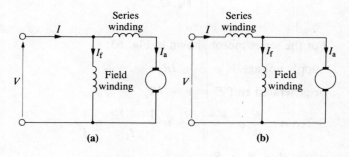

Fig. 65

There are two types of compound motor:

(i) **Cumulative compound,** in which the series winding is so connected that the field due to it assists that due to the shunt winding.

(ii) **Differential compound,** in which the series winding winding is so connected that the field due to it opposes that due to the shunt winding.

Typical compound motor torque and speed characteristics are shown in Fig. 66.

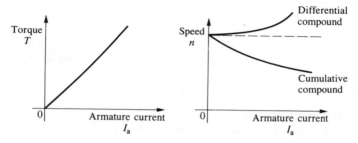

Fig. 66

■ **Separately-excited generator**
For the separately-excited generator shown in Fig. 67:

Terminal voltage $V = E - I_a R_a$

or generated emf $E = V + I_a R_a$

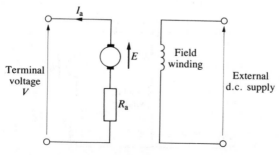

Fig. 67

A typical separately-excited generator **open-circuit characteristic** is shown in Fig. 68(a) and a typical **load characteristic** is shown in Fig. 68(b).

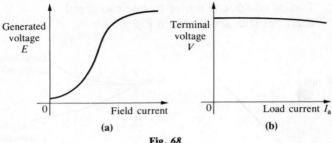

Fig. 68

■ Shunt generator

For the shunt generator shown in Fig. 69:

Terminal voltage $V = E - I_a R_a$

or generated emf $E = V + I_a R_a$

Efficiency of shunt generator,

$$\eta = \left(\frac{VI}{VI + I_a^2 R_a + VI_f + C} \right) \times 100\%$$

where C = sum of iron, friction and windage losses

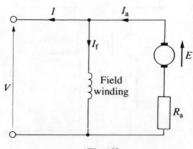

Fig. 69

The **efficiency of a generator** is a maximum when the load is such that:

$$I_a^2 R_a = VI_f + C$$

A typical shunt generator open-circuit characteristic is shown in Fig. 70(a) and a typical load characteristic is shown in Fig. 70(b).

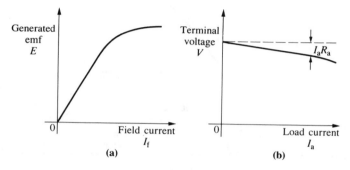

Fig. *70*

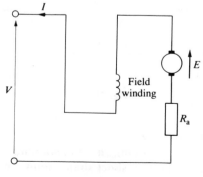

Fig. *71*

■ Series generator

For the series generator shown in Fig. 71:

Terminal voltage $V = E - IR_a$

or generated emf $E = V + IR_a$

A typical load characteristic for a series generator is shown in Fig. 72.

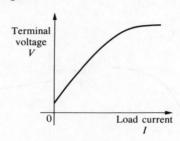

Fig. 72

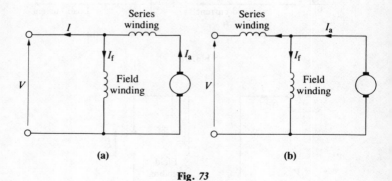

Fig. 73

■ Compound generator

Figure 73(a) shows a **long-shunt compound generator** and Fig. 73(b) shows a **short-shunt compound generator**.

Typical compound generator load characteristics are shown in Fig. 74.

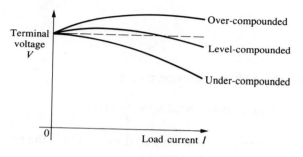

Fig. *74*

INDUCTION MOTORS

For a three-phase induction motor,

Synchronous speed $n_s = \dfrac{f}{p}$ rev/s

where f = frequency of the currents in the stator windings

p = pairs of poles

Slip speed = $(n_s - n_r)$ rev/s   where n_r = rotor speed

per-unit or fractional slip $s = \left(\dfrac{n_s - n_r}{n_s} \right)$

frequency of rotor emf, $f_r = sf$

The **rotor emf generated per phase E_r** is given by:

$E_r = sE_0$ where E_0 = rotor emf generated per phase at standstill

The **rotor impedance per phase Z_r** at slip s is given by:

$Z_r = \sqrt{[R^2 + (sX_0)^2]}$ where R = rotor resistance per phase

X_0 = rotor leakage reactance per phase at standstill

The **rotor current per phase I_r** at slip s is given by:

$$I_r = \frac{sE_0}{\sqrt{[R^2 + (sX_0)^2]}}$$

Slip
$$s = \frac{\text{total rotor } I^2R \text{ loss}}{\text{input power to rotor}} = \frac{2\pi T(n_s - n_r)}{2\pi Tn_s}$$

where T = torque exerted on the rotor by the rotating flux, in newton metres

The **torque T** on the rotor for a given synchronous speed and number of rotor phases is given by:

$$T \, \alpha \, \frac{sE_0^2 R}{R^2 + (sX_0)^2} \, \alpha \, \frac{s\Phi^2 R}{R^2 + (sX_0)^2}$$

For a small value of slip: $T \, \alpha \, \dfrac{s}{R}$

For a large value of slip and low rotor resistance: $T \, \alpha \, \dfrac{R}{s}$

For maximum torque: $R = sX_0$

A.C. BRIDGE NETWORKS

For the general a.c. bridge network shown in Fig. 75:

At balance: $Z_1 Z_3 = Z_2 Z_4$

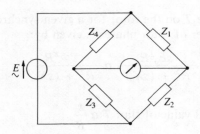

Fig. 75

SOME COMMON A.C. BRIDGE NETWORKS

Type of a.c. bridge network　　　　　*Balance conditions*

(a) **Simple Maxwell bridge**

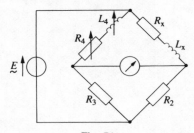

$$R_x = \frac{R_2 R_4}{R_3}$$

$$L_x = \frac{R_2 L_4}{R_3}$$

Fig. 76

| *Type of a.c. bridge network* | *Balance conditions* |

(b) The Hay bridge

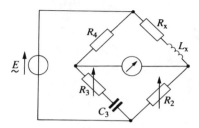

$$R_x = \frac{\omega C_3^2 R_2 R_3 R_4}{(1 + \omega^2 C_3^2 R_3^2)}$$

$$L_x = \frac{C_3 R_2 R_4}{(1 + \omega^2 C_3^2 R_3^2)}$$

Fig. 77

(c) The Owen bridge

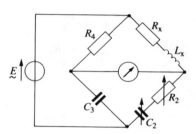

$$R_x = \frac{R_4 C_3}{C_2}$$

$$L_x = R_2 R_4 C_3$$

Fig. 78

(d) The Maxwell–Wien bridge

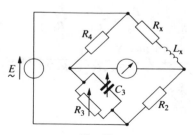

$$R_x = \frac{R_2 R_4}{R_3}$$

$$L_x = C_3 R_2 R_4$$

Fig. 79

Type of a.c. bridge network   *Balance conditions*

(e) **The de Sauty bridge**

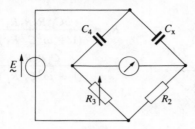

$$C_x = \frac{R_3 C_4}{R_2}$$

Fig. 80

(f) **The Schering bridge**

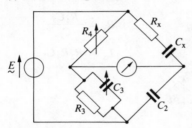

$$R_x = \frac{C_3 R_4}{C_2}$$

$$C_x = \frac{C_2 R_3}{R_4}$$

Fig. 81

(g) **The Wien bridge**

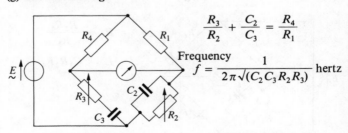

$$\frac{R_3}{R_2} + \frac{C_2}{C_3} = \frac{R_4}{R_1}$$

Frequency
$$f = \frac{1}{2\pi \sqrt{(C_2 C_3 R_2 R_3)}} \text{ hertz}$$

Fig. 82

COMPLEX WAVEFORMS

■ **The general equation for a complex waveform**
The instantaneous value of a complex voltage wave v
acting in a linear circuit may be represented by the
general equation:

$$v = V_0 + V_{1m} \sin(\omega t + \psi_1) + V_{2m} \sin(2\omega t + \psi_2)$$
$$+ \ldots + V_{nm} \sin(n\omega t + \psi_n) \text{ volts} \quad \ldots (1)$$

V_0 represents a d.c. component,
$V_{1m} \sin(\omega t + \psi_1)$ represents the fundamental
component,

where $\qquad V_{1m} =$ peak or maximum value

frequency $f = \dfrac{\omega}{2\pi}$ hertz

$\psi_1 =$ phase angle with respect to
time $t = 0$

Similarly, $V_{2m} \sin(2\omega t + \psi_2)$ and $V_{nm} \sin(n\omega t + \psi_n)$
represent the second and nth harmonic components
respectively.

The instantaneous value of a complex current i may
be represented by the general equation:

$$i = I_0 + I_{1m} \sin(\omega t + \theta_1) + I_{2m} \sin(2\omega t + \theta_2)$$
$$+ \ldots + I_{nm} \sin(n\omega t + \theta_n) \text{ amperes} \quad \ldots (2)$$

If equations (1) and (2) refer to the voltage across and
the current flowing through a given linear circuit,

then the phase angle between the fundamental voltage and current is $\phi_1 = (\psi_1 - \theta_1)$, the phase angle between the second harmonic voltage and current is $\phi_2 = (\psi_2 - \theta_2)$, and so on.

■ RMS value of a complex wave

The rms value of the complex voltage v in equation (1) is given by:

$$V = \sqrt{\left(V_0^2 + \frac{V_{1m}^2 + V_{2m}^2 + \ldots + V_{nm}^2}{2}\right)} \text{ volts}$$

or $V = \sqrt{(V_0^2 + V_1^2 + V_2^2 + \ldots + V_n^2)}$ volts

where $V_1, V_2, \ldots, V_n$ are the rms values of the respective harmonics.

Similarly, the rms value of the complex current i given in equation (2) is given by:

$$I = \sqrt{\left(I_0^2 + \frac{I_{1m}^2 + I_{2m}^2 + \ldots + I_{nm}^2}{2}\right)}$$

or $I = \sqrt{(I_0^2 + I_1^2 + I_2^2 + \ldots + I_n^2)}$

where $I_1, I_2, \ldots, I_n$ are the rms values of the respective harmonics.

■ Mean value of a complex wave

The mean or average value of a complex quantity whose negative half-cycle is similar in shape to its positive half-cycle is given by:

For voltage v given in equation (1),

$$v_{av} = \frac{1}{\pi} \int_0^\pi v \, d(\omega t)$$

and for current i given in equation (2),

$$i_{av} = \frac{1}{\pi} \int_0^{\pi} i \, d(\omega t),$$

each waveform being taken over half a cycle.

■ Form factor of a complex wave

The form factor of a complex waveform whose negative half-cycle is similar in shape to its positive half-cycle is defined as:

$$\text{Form factor} = \frac{\text{rms value of the waveform}}{\text{mean value}}$$

where the mean value is taken over half a cycle.

■ Power associated with complex waves

The average power P supplied for one cycle of the fundamental is given by:

$$P = V_0 I_0 + V_1 I_1 \cos\phi_1 + V_2 I_2 \cos\phi_2 \\ + \ldots + V_n I_n \cos\phi_n \text{ watts}$$

where V_1 and I_1 are the rms values of the fundamental voltage and current respectively

ϕ_1 = phase angle between the fundamental voltage and current, and so on.

Alternatively, if R is the equivalent series resistance of a circuit, then the total power is given by:

$$P = I^2 R \text{ watts} \quad \text{where } I = \text{rms value of complex current } i$$

■ Power factor

When harmonics are present in a waveform, the overall power factor is defined as:

$$\text{Overall power factor} = \frac{\text{total power supplied}}{\begin{array}{c}\text{(total rms voltage} \times \\ \text{total rms current)}\end{array}}$$

i.e. power factor $= \dfrac{(V_0 I_0 + V_1 I_1 \cos\phi_1 + V_2 I_2 \cos\phi_2 + \dots)}{VI}$

■ Harmonics in single-phase circuits

If a complex voltage v represented by

$$v = V_{1m} \sin\omega t + V_{2m} \sin 2\omega t + V_{3m} \sin 3\omega t + \dots \text{ volts}$$

is applied to a single-phase circuit containing:

(a) pure resistance R, then the resulting current i is given by:

$$i = \frac{V_{1m}}{R} \sin\omega t + \frac{V_{2m}}{R} \sin 2\omega t + \frac{V_{3m}}{R} \sin 3\omega t + \dots \text{ amperes}$$

(b) pure inductance L, then the resulting current i is given by:

$$i = \frac{V_{1m}}{\omega L} \sin\left(\omega t - \frac{\pi}{2}\right) + \frac{V_{2m}}{2\omega L} \sin\left(2\omega t - \frac{\pi}{2}\right) + \frac{V_{3m}}{3\omega L} \sin\left(3\omega t - \frac{\pi}{2}\right) + \dots \text{ amperes}$$

(c) pure capacitance C, then the resulting current i is given by:

$$i = V_{1m}(\omega C) \sin\left(\omega t + \frac{\pi}{2}\right)$$

$$+ V_{2m}(2\omega C) \sin\left(2\omega t + \frac{\pi}{2}\right)$$

$$+ V_{3m}(3\omega C) \sin\left(3\omega t + \frac{\pi}{2}\right) + \dots \text{ amperes}$$

■ Resonance due to harmonics

When a circuit resonates at one of the harmonic frequencies of the supply voltage, the effect is called **selective** or **harmonic resonance**.

For resonance at the nth harmonic, $n\omega L = \dfrac{1}{n\omega C}$

FIELD THEORY

The **capacitance C between concentric cylinders** (or coaxial cable) is given by:

$$C = \frac{2\pi\varepsilon_0\varepsilon_r}{\ln\dfrac{b}{a}} \text{ farads/metre}$$

where a = inner conductor radius
b = outer conductor radius
and $\ln = \log_e$

Dielectric stress, $E = \dfrac{V}{r\ln\dfrac{b}{a}}$ volt/metre

where V = core potential
r = conductor radius

$$E_{max} = \frac{V}{a\ln\dfrac{b}{a}}, \; E_{min} = \frac{V}{b\ln\dfrac{b}{a}}$$

The **dimensions of the most economical cable** are given by:

$$a = \frac{V}{E_{max}} \text{ and } b = a\,\mathrm{e}$$

(where e = 2.718 correct to 4 significant figures)

The **capacitance C of an isolated twin line** is given by:

$$C = \frac{\pi \varepsilon_0 \varepsilon_r}{\ln \dfrac{D}{a}} \text{ farads/metre}$$

where D = distance between the centres of the two conductors

a = radius of each conductor

Energy stored in a capacitor, $W = \frac{1}{2}CV^2$ joules

Energy stored per unit volume of dielectric

$$= \tfrac{1}{2}DE = \tfrac{1}{2}\varepsilon_0 \varepsilon_r E^2 = \frac{D^2}{2\varepsilon_0 \varepsilon_r} \text{ joules/metre}^3$$

MAGNETIC FIELDS

The **inductance L of a pair of concentric cylinders** (or coaxial cables) is given by:

$$L = \frac{\mu_0 \mu_r}{2\pi} \left(\tfrac{1}{4} + \ln \frac{b}{a} \right) \text{ henry/metre}$$

where a = inner conductor radius

b = outer conductor radius

111

The **inductance L of an isolated twin line** (i.e. the 'loop inductance') is given by:

$$L = \frac{\mu_0 \mu_r}{\pi} \left(\tfrac{1}{4} + \ln \frac{D}{a} \right) \text{ henry/metre}$$

where D = distance between the centres of the two conductors
a = radius of each conductor

Energy stored in a non-magnetic medium

$$= \tfrac{1}{2}BH = \tfrac{1}{2}\mu_0 H^2 = \frac{B^2}{2\mu_0} \text{ joules/metre}^3$$

Energy stored in an inductor, $W = \tfrac{1}{2}LI^2$ joules.

LOGARITHMIC RATIOS

The ratio of two powers P_1 and P_2 may be expressed in logarithmic form.

$$\text{Power ratio in decibels} = 10 \lg \frac{P_2}{P_1}$$

where P_1 = input power to a system
P_2 = output power to a system
and $\lg = \log_{10}$

or power ratio in nepers $= \frac{1}{2} \ln \frac{P_2}{P_1}$

If P_1 and P_2 refer to power developed in two equal resistors, then:

$$\text{Ratio in decibels} = 20 \lg \frac{V_2}{V_1} = 20 \lg \frac{I_2}{I_1}$$

or ratio in nepers $= \ln \frac{V_2}{V_1} = \ln \frac{I_2}{I_1}$

ATTENUATORS

An **attenuator** is a network for introducing a specified loss between a signal source and a matched load without upsetting the impedance relationship necessary for matching.

Attenuation is a reduction in the magnitude of a voltage or current due to its transmission over a line or through an attenuator.

Networks in which electrical energy is fed in at one pair of terminals and taken out at a second pair of terminals are called **two-port networks**.

For any passive two-port network it is found that a particular value of load impedance can always be determined which will produce an input impedance having the same value as the load impedance. This is called the **iterative impedance** for an asymmetrical network and its value depends on which pair of terminals is taken to be the input and which is the output; there are thus two values of iterative impedance, one for each direction. For a symmetrical network, the two iterative impedances are equal and this value is called the **characteristic impedance**.

For the **symmetrical T-attenuator** shown in Fig. 83, the characteristic impedance R_0 is given by:

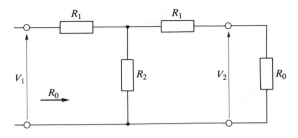

Fig. 83

$$R_0 = \sqrt{(R_1^2 + 2R_1R_2)}$$
$$\text{or} \quad R_0 = \sqrt{(R_{oc}R_{sc})}$$

where R_{oc} = input resistance when the output is open-circuited

R_{sc} = input resistance when the output is short-circuited

If the characteristic impedance R_0 and the attenuation $N \left(= \dfrac{V_1}{V_2} \right)$ are known for the T-attenuator of Fig. 83, then:

$$R_1 = R_0 \left(\frac{N-1}{N+1} \right) \text{ and } R_2 = R_0 \left(\frac{2N}{N^2 - 1} \right)$$

For the **symmetrical π-attenuator** shown in Fig. 84, the characteristic impedance R_0 is given by:

$$R_0 = \sqrt{\left(\frac{R_1R_2^2}{R_1 + 2R_2} \right)}$$
$$\text{or} \quad R_0 = \sqrt{(R_{oc}R_{sc})}$$

If the characteristic impedance R_0 and the attenuation

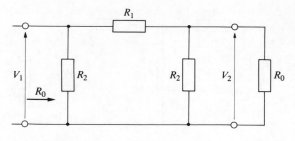

Fig. *84*

$N\left(=\dfrac{V_1}{V_2}\right)$ are known for the π-attenuator of Fig. 84, then:

$$R_1 = R_0\left(\frac{N^2 - 1}{2N}\right) \text{ and } R_2 = R_0\left(\frac{N + 1}{N - 1}\right)$$

For an attenuator the **insertion loss ratio** A_L is given by:

$$A_L = \frac{\text{(voltage across load when connected directly to a generator)}}{\text{(voltage across load when a two-port network is connected)}}$$

For a network terminated in its characteristic impedance,

$$\text{Insertion loss} = 20 \lg \frac{V_1}{V_2} \text{ dB or } 20 \lg \frac{I_1}{I_2} \text{ dB}$$

An **image impedance** is the impedance which, when connected to the terminals of a network, equals the impedance presented to it at the opposite terminals.

For example, with a two-port network, if the output is terminated with an impedance Z_2 when the input impedance is Z_1, and if the input is terminated with an impedance Z_1 when the output impedance is Z_2, then Z_1 and Z_2 are called the image impedances. An asymmetrical network is correctly terminated when it is terminated in its image impedance.

For the **L-section attenuator** shown in Fig. 85:

$$R_1 = \sqrt{[R_{OA}(R_{OA} - R_{OB})]}$$

$$\text{and} \quad R_2 = \sqrt{\left(\frac{R_{OA} R_{OB}^2}{R_{OA} - R_{OB}}\right)}$$

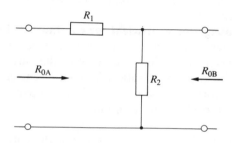

Fig. 85

□ *Two-port networks in cascade*
The overall attenuation a of $(n - 1)$ matched sections connected in cascade is given by:

$$a = a_1 + a_2 + \ldots + a_{n-1}$$

where $a = 20 \lg \dfrac{V_1}{V_n}$ dB

a_1 = attenuation of first section in decibels, i.e.

$$a_1 = 20 \lg \frac{V_1}{V_2}, \text{ and so on.}$$

117

TRANSMISSION LINES

A **transmission line** is a system of conductors connecting one point to another and along which electromagnetic energy can be transmitted.

The four **primary constants** of a transmission line are resistance R, inductance L, capacitance C and conductance G.

The **propagation coefficient (or constant)** Υ, for a current wave at a specified frequency and transmitted along a transmission line, is defined as the natural logarithm of the phasor ratio of the input current to the output current at points unit distance apart. The propagation coefficient has no units.

If I_S is the sending-end current and I_1 is the current 1 m along a line,

$$\text{then } \frac{I_S}{I_1} = e^{\Upsilon} \text{ from which } \Upsilon = \ln \frac{I_S}{I_1}$$

The propagation coefficient is a complex quantity given by:

$$\Upsilon = \alpha + j\beta \quad \text{where } \alpha = \text{attenuation coefficient in nepers/metre}$$
$$\beta = \text{phase-change coefficient in radians/metre}$$

The receiving end current I_R is given by:

$$I_R = I_S e^{-nY} = I_S e^{-n\alpha} \angle -n\beta \text{ amperes}$$

where I_S = sending-end current in amperes
   n = length of line in metres

The receiving-end voltage V_R is given by:

$$V_R = V_S e^{-nY} = V_S e^{-n\alpha} \angle -n\beta \text{ volts}$$

where V_S = sending-end voltage in volts

In a transmission line there will be a time, and thus phase difference, between the generator input voltage and the voltage at any point on the line. The phase-change coefficient β is given by:

$$\beta = \omega \sqrt{(LC)} \text{ radians/metre}$$

where L and C are the inductance and capacitance per metre of the line (when $\omega L \gg R$ and $\omega C \gg G$)
The wavelength λ on a line is the distance between a given point and the next point along the line at which the voltage has the same phase, the initial point leading the latter point by 2π radians.

$$\lambda = \frac{2\pi}{\beta} \text{ metres}$$

The velocity of propagation u is given by:

$$u = f\lambda = \frac{\omega}{\beta} \quad \text{where } f = \text{frequency in hertz}$$

The four **secondary line constants** of a transmission line are the characteristic impedance Z_0, propagation coefficient Y, attenuation coefficient α and phase-shift coefficient β.

The characteristic impedance Z_0 is given by:

$$Z_0 = \sqrt{(Z_{oc} Z_{sc})} \text{ ohms} \quad \text{where } Z_{oc} = \text{open-circuit}$$
$$\text{impedance}$$
$$Z_{sc} = \text{short-circuit}$$
$$\text{impedance}$$

The characteristic impedance may also be expressed in terms of the primary constants,

$$\text{i.e. } Z_0 = \sqrt{\left(\frac{R + j\omega L}{G + j\omega C}\right)} \text{ ohms}$$

If $\omega L \gg R$ and $\omega C \gg G$, $\quad Z_0 = \sqrt{\left(\frac{L}{C}\right)} \text{ ohms}$

The propagation coefficient γ may be expressed in terms of the primary constants,

$$\text{i.e. } \gamma = \sqrt{[(R + j\omega L)(G + j\omega C)]}$$

If $\omega L \gg R$ and $\omega C \gg G$, $\quad \gamma = j\omega\sqrt{(LC)} = j\beta$
$$(\text{i.e. } \alpha = 0)$$

If the waveform at the receiving end of a transmission line is not the same shape as the waveform at the sending end, **distortion** is said to have occurred.

In a transmission line, if $LG = CR$ it is possible to provide a termination equal to the characteristic impedance at all frequencies. This condition is the most appropriate for the design of a transmission line since under this condition no distortion is introduced.

Increasing the line inductance L, called **loading**, is needed to achieve this condition (although this is difficult to attain in practice).

■ Wave reflection and reflection coefficient

When the terminating impedance of a transmission line does not have the same value as the characteristic impedance of the line, the transmission line is said to have a **'mis-matched load'**. When the input and output impedances of a transmission line are equal, the line is said to be **'correctly terminated'**.

The forward travelling wave moving from the source to the load is called the **incident wave**. With a mismatched load the termination will be able to absorb only a part of the energy of the incident wave, the remainder being forced to return back along the line towards the source. This latter wave is called the **reflected wave**. The ratio of the reflected current to the incident current is called the **reflection coefficient** ρ.

$$\rho = \frac{I_r}{I_i} = \frac{Z_0 - Z_R}{Z_0 + Z_R} = -\frac{V_r}{V_i}$$

where I_i = incident or sending-end current
V_i = incident or sending-end voltage
I_r = reflected current
V_r = reflected voltage
Z_0 = characteristic impedance
Z_R = impedance of termination

When $Z_R = Z_0$, $\rho = 0$ and there is no reflection.

■ Standing waves and standing wave ratio

Whenever two waves of the same frequency and amplitude travelling in opposite directions are superimposed on each other, interference takes place between the two waves and a **standing** or **stationary wave** is produced.

The total current or voltage at any point on a transmission line is the phasor sum of the incident and reflected waves.

The points at which the total current or voltage waves are at a minimum are called **nodes**. Those points on the wave where the total current or voltage waves are at a maximum are called **antinodes**.

The **standing wave ratio** s on a transmission line is given by:

$$s = \frac{I_{max}}{I_{min}} = \frac{I_i + I_r}{I_i - I_r}$$

Also $\dfrac{I_r}{I_i} = |\rho| = \dfrac{s - 1}{s + 1}$

or $s = \dfrac{1 + |\rho|}{1 - |\rho|}$

where $|\rho|$ = magnitude of reflection coefficient

and $\dfrac{P_r}{P_t} = \left(\dfrac{s - 1}{s + 1}\right)^2$

where P_r = reflected power
P_t = power absorbed in the termination

COMMON ELECTRICAL CIRCUIT DIAGRAM SYMBOLS (BS3939, 1985)

DESCRIPTION	SYMBOL	DESCRIPTION	SYMBOL
Conductor or group of conductors		Fuse, general symbol	
Three conductors: single-line representation		Fuse with the supply side indicated by a thick line	
Junction of conductors	OR	Resistor, general symbol	
Double junction of conductors	OR	Alternative general symbol for resistor	
Primary cell or accumulator. The longer line represents the positive pole, the short line the negative pole. (The short line may be thickened for emphasis)		Variable or adjustable resistor	
		Resistor with pre-set adjustment	
Battery of accumulators or primary cells		Potentiometer with sliding contact	
Alternative symbol for battery of accumulators or primary cells		Resistor with sliding contact	
		Voltage dependent resistor (Note: U may be replaced by V)	U
Photovoltaic cell		Temperature dependent resistor (Note: Θ may be replaced by $t°$.)	Θ
Positive polarity	$+$	Light-dependent resistor	
Negative polarity	$-$		
Earth or ground, general symbol		Inductor, coil, winding or choke	
Frame or chassis		Variable inductor, coil, winding or choke	
Direct current		Inductor with pre-set adjustment	
Alternating current	$\sim$		
Three phase delta-connected winding	$\triangle$	Inductor with magnetic core	
Three phase star-connected winding	$\curlyvee$	Inductor with gap in magnetic core	
Switch, general symbol	OR		

DESCRIPTION	SYMBOL	DESCRIPTION	SYMBOL
Transformer with two windings	OR air core / magnetic core	Bell	
		Single-stroke bell	
		Buzzer	
Auto-transformer	OR	Ideal current source	
		Ideal voltage source	
Current or pulse transformer	OR	Indicating instrument, general symbol	
Capacitor, general symbol		Voltmeter	(V)
Variable or adjustable capacitor		Ammeter	(A)
Capacitor with pre-set adjustment		Wattmeter	(W)
Polarised capacitor, for example, electrolytic		Ohmmeter	(Ω)
Microphone, general symbol		Power-factor meter	(cosφ)
Earphone, general symbol		Phase meter	(φ)
		Frequency meter	(Hz)
Loudspeaker, general symbol		Thermometer or pyrometer (Note: Θ may be replaced by $t°$)	(Θ)
Antenna, general symbol		Tachometer	(n)
		Oscilloscope	
Lamp, general symbol	$\otimes$	Galvanometer	

124

DESCRIPTION	SYMBOL	DESCRIPTION	SYMBOL
Thermocouple, shown with polarity symbols		Positive-going pulse	
Thermocouple with direct indication of polarity, the negative pole being represented by the thick line.		Negative-going pulse	
		Pulse of alternating current	
Clock, general symbol Secondary clock		Positive-going step function	
		Negative-going step function	
Master clock		Saw-tooth	
Clock with switch			
Recording instrument, general symbol		Amplifier, general symbol	
Recording wattmeter	W	Alternative general symbol for amplifier	
Integrating instrument or Energy meter, general symbol		High-gain differential amplifier (operational amplifier)	
Watt-hour meter	Wh		
Motor	M	Inverting amplifier with an amplification of 1. $u = -1.a$	
D.C. motor	M		
A.C. motor	M	Summing amplifier $u = -10(0.1a + 0.2b + 0.5c)$ $= -(a + 2b + 5c)$	
Generator	G		
D.C. Generator	G	Integrating amplifier (integrator) $u = -80 \int_{\cdot}^{t} (2a + 3b + 7c) dt$	
A.C. Generator	G		
Machine capable of use as a generator or motor	MG	Differentiating amplifier (differentiator) $u = 5 \frac{d}{dt}(a - 4b)$	
Synchronous motor	MS		
Synchronous generator	GS		

DESCRIPTION	SYMBOL	DESCRIPTION	SYMBOL
Attentuator, fixed loss (Pad)	dB	Photodiode	
Filter, general symbol		Diode where use is made of its temperature dependence (Note: Θ may be replaced by $t°$)	θ
		Variable capacitance diode (varactor)	
High-pass filter		Tunnel diode	
Low-pass filter		Breakdown diode, unidirectional (for example, Zener diode)	
Band-pass filter		Breakdown diode, bi-directional	
		Bi-directional diode. Diac	
Band-stop filter		Reverse blocking diode thyristor	
Rectifying junction	OR	Reverse conducting diode thyristor	
		Bi-directional diode thyristor	
Rectifier		Triode thyristor, type unspecified	
Inverter		Reverse blocking triode thyristor, N-gate (anode-side controlled)	
Rectifier in full wave (bridge) connection		Reverse blocking triode thyristor, P-gate (cathode-side controlled)	
Semiconductor diode, general symbol		Turn-off triode thyristor, gate not specified	
		Reverse blocking thyristor tetrode type	
Light-emitting diode, general symbol		Bi-directional triode thyristor. Triac	

DESCRIPTION	SYMBOL	DESCRIPTION	SYMBOL
PNP transistor		IGFET enhancement type, single gate, N-type channel with substrate internally connected to source	
Phototransistor, PNP type		IGFET depletion type, single gate, N-type channel without substrate connection	
NPN transistor with collector connected to the envelope		IGFET depletion type, single gate, P-type channel without substrate connection	
NPN avalanche transistor		IGFET depletion type with two gates, N-type channel with substrate connection brought out	
Unijunction transistor with P-type base			
Unijunction transistor with N-type base		Optical coupling device	
Junction field effect transistor with N-type channel		Conductive coating on internal surface of envelope	
Junction field effect transistor with P-type channel		Hot cathode, indirectly heated	
Insulated gate field effect transistor (abridged IGFET) enchancement type, single gate, P-type channel without substrate connection		Alternative symbol for hot cathode, indirectly heated	
IGFET enhancement type, single gate, N-type channel without substrate connection		Hot cathode, directly heated. Heater for hot cathode, indirectly heated. Heater for thermocouple	
IGFET enhancement type, single gate, P-type channel with substrate connection brought out			

COMMON LOGIC SYMBOLS		
DESCRIPTION	BRITISH	AMERICAN
AND $f = x.y.z$		
OR $f = x+y+z$		
NOT, NEGATOR OR INVERTOR		
NAND $f = \overline{x.y.z}$		
NOR $f = \overline{x+y+z}$		
EXCLUSIVE OR $f = x \oplus y$		
EXCLUSIVE NOR $f = \overline{x \oplus y}$		

128

RESISTOR COLOUR CODING AND OHMIC VALUES

COLOUR CODE FOR FIXED RESISTORS

COLOUR	SIGNIFICANT FIGURES	MULTIPLIER	TOLERANCE
Silver	—	10^{-2}	$\pm 10\%$
Gold	—	10^{-1}	$\pm 5\%$
Black	0	1	—
Brown	1	10	$\pm 1\%$
Red	2	10^2	$\pm 2\%$
Orange	3	10^3	—
Yellow	4	10^4	—
Green	5	10^5	$\pm 0.5\%$
Blue	6	10^6	$\pm 0.25\%$
Violet	7	10^7	$\pm 0.1\%$
Grey	8	10^8	—
White	9	10^9	—
None	—	—	$\pm 20\%$

Thus, for a **four-band fixed resistor** (i.e. resistance values with two significant figures):

> yellow–violet–orange–red indicates 47 kΩ with a tolerance of $\pm 2\%$,
> orange–orange–silver–brown indicates 0.33 Ω with a tolerance of $\pm 1\%$,

and brown–black–brown indicates 100 Ω with a tolerance of $\pm 20\%$.

(Note that the first band is the one nearest the end of the resistor.)

For a **five-band fixed resistor** (i.e. resistance values with three significant figures):

> red—yellow—white—orange—brown indicates 249 kΩ with a tolerance of ± 1%

(Note that the fifth band is 1.5 to 2 times wider than the other bands.)

LETTER AND DIGIT CODE FOR RESISTORS

RESISTANCE VALUE	MARKED AS:
0.47 Ω	R47
1 Ω	1R0
4.7 Ω	4R7
47 Ω	47R
100 Ω	100R
1 kΩ	1K0
10 kΩ	10K
10 MΩ	10M

Tolerance is indicated as follows:

F = ± 1%, G = ± 2%, J = ± 5%,
K = ± 10% and M = ± 20%

Thus, for example, R33M = 0.33 Ω ± 20%
4R7K = 4.7 Ω ± 10%
390RJ = 390 Ω ± 5%
6K8F = 6.8 kΩ ± 1%
68KK = 68 kΩ ± 10%
4M7M = 4.7 MΩ ± 20%

Index